A Lakota Approach to Biodynamics

A Lakota Approach to Biodynamics

— Taking Life Seriously —

Devon Strong

Edited by
Susan C. Strong and Dawn van Buuren

Lindisfarne | 2017

In Memory of Devon Strong

And while I stood there I saw more than I can tell and I understood more than I saw; for I was seeing in a sacred manner the shapes of all things in the spirit, and the shape of all shapes as they must live together like one being.
    - John G. Neihardt, *Black Elk Speaks*

The plants and animals on Earth...cannot be understood in isolation...it is nonsense to seek within the compass needle itself, the reason why it always points to the north...Just as we have to look at the whole Earth when we want to explain how a compass needle behaves, so must we also consult the whole universe when it comes to understanding plants.
    - Rudolf Steiner, *Spiritual Foundations for the Renewal of Agriculture* (Lecture 6)

Published by Lindisfarne Books,
an imprint of Anthroposophic Press, Inc.
610 Main Street
Great Barrington, Massachusetts 01230

www.steinerbooks.org

Copyright © 2017 by Zachary Strong van Buuren
All rights reserved.

No part of this book may be reproduced in any form without written permission from the publisher, except for brief quotations embodied in critical articles for review.

Print ISBN: 978-1-58420-973-7
e-book ISBN: 978-1-58420-974-4

Printed in the USA

Contents

Introduction......xi

I. Taking Life Seriously by Devon Strong
(unfinished 2015 Book Draft)

Preface......3
Chapter One: *Animals and Farms*......5
Chapter Two: *Animal Consciousness and Group Soul*......9
Chapter Three: *Animal Consciousness and the Individual*......15
Chapter Four: *The Spiritual Sheath of Animals*......20
Chapter Five: *Buffalo, Biodynamics, and Native Ceremony*......25
Chapter Six: *Birth and Death in the Group Soul*......36
Chapter Seven: *Blood and Guts*......44

II. Poetry by Devon Strong
Buffalo and the Blade......53
Coming Together......57

III. What Others Say
Bison and the Sacred Hoop of Life by Catherine Preus......61
Letter from Jean-Michel Florin......66
My Encounter with Devon Strong by Uli Johannes Köenig......68
An Unofficial Farm Report by Anke van Leewen......73

IV. Obituaries
Buffalo Rancher Found Dead in Knife Accident, by John Darling......87
Some Thoughts on Devon Strong by Jim Fullmer......90

V. Aftermath
The Strong Buffalo Story by Craig Strong......95

Afterword......105

Appendix: Previously Published Articles by Devon Strong
My Experience Raising Buffalo......111
Taking Life Seriously......120
In Partnership with Animals......128

Acknowledgements......131

Introduction

WITH THE TRAGIC and unexpected death of Devon Strong in the fall of 2015, his family was left with many tasks. Taking charge of Four Eagles Farm, finding new homes for all of his animals (17 buffalo, 28 sheep, 3 horses, and a dog), and planning a memorial for him were among the many things we had to do first and fast.

The book you hold in your hands was to be another of Devon's dozens of projects. After he was invited twice to speak about his Lakota version of biodynamics at the International Biodynamic Agriculture Conference at the Goetheanum in Dornach, Switzerland, he had been offered a book contract by SteinerBooks. After his death, his father and I contacted them to see if they were still interested. They were, and because of the unfinished nature of his book manuscript, we all agreed that adding a compilation of other relevant articles would be the best way to go forward.

This book contains the unfinished but edited manuscript of his book, several of his poems, plus letters and articles by those who knew him and his work. It includes an essay by Anke van Leewen, who visited Devon's farm for two weeks in September 2015 on behalf of the Section for Agriculture at the Goetheanum, as part of a team studying Devon's biodynamic preps. There are also two obituaries that speak to different aspects of Devon's life and work, followed by his brother Craig Strong's "Buffalo Story." That's the tale of how we, his family, with help from his Lakota community and so many others, sent Devon's beloved buffalo to Knife Chief Buffalo Nation. Knife Chief Buffalo Nation is a nonprofit in South Dakota, on the Pine Ridge Reservation, working to revive Lakota pride and ways. Buffalo are an intrinsic part of their mission. Finally, we have placed articles Devon had already published in an Appendix at the end of the book.

In the Afterword, we've included some thoughts about the potential for others to carry on Devon's pioneering work in a variety of ways.

It is appropriate to say a few words about the principles Dawn van Buuren and I followed in editing Devon's unfinished book manuscript. Devon's worldview was that of the Lakota community who embraced him. That approach is deeply reverent with respect to spirit in nature and understands life as a circle and as a cycle. Very early in life Devon had left behind the reductive linear perspective of modern western civilization, now so clearly being revealed as toxic to all life on this planet. Starting his farm career with organic gardening, Devon added buffalo to the mix and learned about how to do Lakota ceremony with them through revelations in the sweat lodge. (Devon had tried to learn of these from modern

Lakota people and other tribal peoples first but was unable to find any memories left of these ancient ceremonies. Too much trauma from the deliberate destruction of tribal culture by whites had erased them.)

Looking for ways to integrate organic farming with his spiritual work with the buffalo, he found biodynamic farming, and for him the circle of spiritual farming was then complete.

In writing about his work with the buffalo and biodynamic agriculture, he was clearly hoping to reach the world with a message about the urgent need to revive our awareness of the sacredness of all life and to recall our place within it, connected to the whole of nature.

Because of all of this, Dawn and I simply edited Devon's unfinished book manuscript for clarity from one sentence to the next and for all of the other usual editorial concerns—punctuation, grammar, spelling, and so on. What we did not do was alter the relationship of his chapters to each other or try to reduce repetition within or between them. There is a lot of repetition in Devon's writing, in his articles and in the book draft itself, but it is a spiraling kind of repetition—each iteration of the ideas he is trying to convey has a way of becoming more detailed and full of valuable information than the last. His spiraling impulse was also a way to reflect his awareness of the interconnectedness of all things. We felt it was very important to leave that kind of circling intact in every way possible.

There is also some repetition in the various articles written by others about Devon that we included. This was not only inevitable, but also a plus for providing a variety of perspectives on who Devon was and about his work.

Finally, who are "we" of the team creating this book? We are Devon's father, Richard Strong and myself, Susan Strong; Craig Strong, his brother; Zachary van Buuren, his son, Dawn van Buuren, his daughter-in-law; and Anke van Leewen, Devon's biodynamic colleague and close friend. Three of us have published books of our own, and so we were familiar with the process of creating a book. Both Dawn and I are professional editors of one kind or another. We took on the job of creating this book together as a memorial to Devon. We hoped to preserve and publicize his two messages. The first is, as I said above, about the need for our own lives to be experienced within the web of life again. The second is the story of his highly creative contributions to the development of biodynamic agriculture, in service to the future of life on our planet.

We would also like to thank all of the other authors, periodicals or online blog editors who gave us permission to use their materials, or provided leads to other contacts and materials. A detailed list of these acknowledgements forms the final section of this book.

As Devon would say in closing:

Health and help,

Susan C. Strong

I
TAKING LIFE SERIOUSLY
by Devon Strong

(*The unfinished 2015 book draft*)

Preface

WHY DO WE have animals on the farm? Certainly one reason is for producing income for the farm. To do this, the animals must reproduce and live healthy lives. But in biodynamics, we consider how to integrate the other senses, including everything from the subconscious to the superconscious. There are more senses than the typical number allotted by traditional science.

I hope that I can write clearly here about the way of taking life with the same care that biodynamic agriculture has for restoring the life forces. My experience of working with many animals, plus my studies of the Lakota way of life, combined with the ideas put forth in Rudolf Steiner's Agricultural Course, have led me to certain conclusions. I have found that the group soul of animals, as expressed in their flocks and herds, reflects differences in the way we humans care for the different archetypes of animals, both as farmers and as human beings. These differences have important effects on our lives and health, as well as on the lives of our animals, with influences going far beyond what is normally understood in conventional agriculture today.

Chapter One
Animals and Farms

Modern farms don't raise family units: they put bulls in one pasture, mother cows separated from feedlot calves, heifers, fat steers—all tuned to the music of selling products. Feedlots and dairies are the epitome of modern farming, but also the source of concentrating negative life forces which can create more problems than we can solve. The farmer chooses who lives and who dies; natural selection turns into his/her choice. Keeping sick lambs alive, or treating injured animals seems to be normal on most farms today. A healthy farm should not have that problem. Accidents are preventable. Health is the highest achievement, and choosing to raise animals requires the knowledge and experience to prevent problems before they happen. That is what we, as farmers, try to accomplish in building infrastructure that can help hold safe these lives we are caretaking, while the livestock do the work of taking care of our farms. The animals' influence pervades the fabric of every place they can access.

On my farm, I select mothers that will raise their own young; ones that have problems or won't take care of babies have no place. I also make minimal efforts to recover sick animals. Injuries are usually my own fault for having a dangerous situation not repaired in time.

We can keep a farm healthy by recycling, revitalizing the soil with preps and compost, renewing the blood lines, harvesting the fruits of our labor, and culling the sick and very old animals. Recognizing disease and knowing remedies is our work as caretakers. The herd/flock also recognizes a disease in an individual, and our observation makes us aware that the group is behaving differently towards an individual. Picking out that individual and treating it is our responsibility. Spending the time and money to repair what is wrong requires a choice to save the individual or to take care of the group. In many cases, it is the group that needs care, at the cost of the individual life. If the healing is not extensive or expensive, the individual can return, but in many cases it is death that results, also known as culling and selling.

In modern industrial agriculture, the need to cull animals is the source of a lot of modern animal welfare troubles, since culling is a liability to industrial processing. That model is set up for healthy young animals, who are the majority of the harvest from the farms. The horrible photos of animals we see being tortured in industrial settings usually come from cleaning up the mess of culled animals blocking the production lines. Workers in the industrial agriculture system have lost contact with the individual (and even their own humanity) in working with industrial scale production. If all animal lives mattered, we wouldn't have food produced in this manner; there would be recognition of animals being at risk of injury or sickness and a protocol for handling these "at risk" animals. They would be sorted out before they get into the "main line" of the meat processing machinery. There is a grading system for meat, and this is where improvements for animals and processing plants could be made. It would require having people who understand animals in their own hearts, and who can read the symptoms of disease, even as the animals themselves realize what is happening.

Then, of course, there are accidents where anyone can be hurt or injured in the situation of moving animals in confinement. In that case confinement is the problem, and it is such a common occurrence in humanity that we see it as normal, even when too many are confined, and we have "traffic jams" due to a restricted flow.

When bison have a sense of confinement, they respond by going berserk—willing to kill themselves just to get out; it is much like claustrophobia, and a reminder of what humanity is increasingly putting up with, which some animals are not willing to do.

We could learn a lesson about life from this example, as so many forms of nature are trying to let us know that what humans are doing is not natural or healthy. These actions cause a disease that pervades the human psyche and makes us susceptible to many more such maladies. I once spent a couple of hours attempting to unload a buffalo bull at an abattoir when it was required to have it inspected to sell the meat, before that law was abolished by the industry in the 1990s as unworkable for bison. The bull did not want to get out of the trailer, and when he was forced out, he did more than enough damage to destroy the corrals all around it. When they were no longer serviceable, we let him back into the trailer, where he was killed and dragged into the facility by a side door the next day. These were some of the extraordinary efforts that had to be made to serve the USDA inspection laws about handling live animals before slaughter. As such laws clearly did not serve the interest of health for the animals or the industry facilities, they were changed. This gives me hope that the laws about humane treatment of animals can be adapted to caring for the animals in mindful ways as well.

Most farms in America are run as family corporations; they are like multi-generational worker bees in the hive of production. But it's not quite as bad as it sounds because the people are preserving their way of life amidst political and government regulations that would make it very hard to live on the land otherwise. Most farms in my area (NW USA) are larger acreages (600-6000 acres), owned for over 50 years, and have either been subdivided into large family acreages that work cooperatively, sharing equipment and resources, or as mostly inter-generational independent farms passing on a dwindling over-regulated land base to the next generation. However, this also creates more unhealthy conditions for animals and people alike, in both the physical and spiritual sense.

CHAPTER TWO
Animal Consciousness and Group Soul

THE EXAMPLE OF HONEYBEES is a nice introduction to animal archetypes, as it refers so directly to the relationship between how they live as a collective, and how the group soul aspect relates to the interaction of their spiritual work with their physical presence. This physical presence can be seen in their work with nectar and pollen as food and drink. But in the spiritual sense, their work is a relationship with the landscape that is much more intentional as individuals visit the flowers and plants. The timing and interspecies relationship of plants and insects has the added dimension of inspiring the fire and air spirits (salamanders and fairies or sylphs) that are also part of the earth sense of fertilization in the plants. So, it is the plants as much as the bees that are spiritually involved within this relationship. As the plants mature and flower, they bring their expression into focus and out into the spheres where the plant spirit is released, which in plants is so different from the animal/insect group aspect.

The plant spirit is a more primal and singular expression of species; each individual's maturation brings a climax of such expression into the star nation. As each flower reflects that star impulse and the burst of energy attracts the pollinator's attention, the introspective bud leads to a seeding explosion—a microcosm of universal expansion. This is elementary compared to the animal group interaction of the collective intelligence of bees. Even as each individual collects the nectar and pollen from each flower and then relates that information to the rest of the hive, the spiritual representatives in the form of fairies follow the bees' progress through the fields. At the same time, the gnomes hitch a ride in the pollen itself, as it transfers from flower to flower. The fire of the stars is carried within the flowers themselves, that reflection of the sun's face in each individual reaching back to itself out in the cosmos. That expression is what attracts the insects energetically to the fact that the flower is at that point of ecstasy and has the essence of spirituality to offer as reward for the transfer of pollen.

The animal version of archetype is more sedate and earthy in most cases, especially in snakes and fish—who in life are fairly quick. They hold the space of slow contemplation in spirit, a deep earth nature, like stone. The gnomes are in their bones, but the fire of quickness is literally in the salamander when they bite.

For myself, I am most familiar with my farm animals (sheep, goats, horses, chickens, cows, and bison) and the nature of wild animals close to my environs: wild coyote, mountain lions, bears, squirrels, birds, deer, elk, hawks, eagles, vultures, magpies, and crows. I have a passing acquaintance with geese, ducks, doves, cranes, burrowing owls, bats, toads, frogs, snakes, crickets, grasshoppers, flies, bees, beetles, mice, rats, worms, and gophers. These are all a part of my life on the farm along with dogs, cats, chickens, sheep, buffalo, and horses.

Now, the different archetypes of the animals in biodynamics are related to their group soul expression of physical relationship, as in a chicken's "spark of life" expression of interest in every small thing within their level of perception. This is quite different from the cow, who is centered on the internal world of digestion, related to the deep earth sense. The bison, while seeming to have a greater awareness than a cow, is just more focused in the body than in the head. The bison lives a wilder lifestyle, centered on the family orientation of raising the next generation, with the instinctual tribal intuition of a matriarchal society as the physical reality. The bison's spiritual aspect is based on the same principle as the cow, occupying the same niche as the bovine in Europe. The bison fulfills the same needs of the environment for an archetypal renewal of earth in North America, Mexico to Canada, and coast to coast.

My sheep are another good expression of group soul as they have a direct link to each other in their flocking tendencies. That is also protection from predators, when they form a group to save the individual from selection, a reflection of the group soul mentality. It was in sheep that I first noticed the difference in killing techniques having an effect on the group. When I shear sheep in the springtime, I have noticed that a few flocks were extremely agitated when I selected an animal for shearing, and the individual would fight as if for its life. While normally I could start the shearing easily, these flocks had recently been visited by the butcher who had shot a few of the animals in the process of harvesting them. The trauma to their group soul was evident. I compared this to what happens when I do a taking of life ceremony using a blade to cut the throat. The rest of the group actually watches sometimes, and they often bed down on the site with the fresh blood. It is relatively quiet and completely non-traumatic. When I first offered to do it with a knife at a farm where I was picking up a lamb to kill, the owner offered to shoot it for me, and in the process he shot it a couple of times in the head, causing a huge uproar in the group without killing it. When I got finally got hold of it and cut its throat with a blade, there was a dramatic difference in the group's behavior.

As this was many years before I started my ceremonial practice or had buffalo, it became clear to me later on when I reflected on it as a biodynamic farmer that the trauma is more with the group than the individual. In the case of taking a selection of animals to the abattoir, it removes them from the consciousness of the flock/herd when they are killed "humanely" by knocking them unconscious by whatever means. The use of a bullet to the brain is used to stun them, but it doesn't necessarily kill them, while the letting of blood is the real killer. I have heard of instances where an animal recovers consciousness in the process of butchering when the blood has not been let, and it is not a pretty sight. This is not a natural process.

In nature, there is an intricate web of archetypes, expressed through the different orders of animals. It is reflected in the native ways of placing animals as archetypes in the cosmology of ceremony: eagles above, the mole in the earth, the elk, bear, buffalo, and wolf in the cardinal directions—sometimes the horse is also brought in. Different tribes had different animals representing the directions: beaver and moose in the north or in the east, the deer, panther, or fox in the south, the snake, the jaguar, as well as the parrot, with the eagle. I am sure that with over 500 different tribes in the Americas, every region had its favorite animals represented in their proper place in the local web of nature.

That web of archetypes represents the importance of siting animals in their place and in the cosmology of the landscape. It also relates to the animals on a farm who are so intimately connected with its development as an organism in its own right. Connecting the farm animals' archetype to the farm is much like the native honoring of animals in the landscape by their ceremonies. Making that connection through ceremonies works with the group soul of all animals, as well as with the individual's. Recognizing that it is still possible, the farmer can make that ceremonial connection between his animals and the farm today too.

In the modern context, I think most people recognize the bison's place—holding earth energy as obvious. I recognized the archetype as the same niche that cattle represent in European agriculture, especially when the description of the archetype more closely resembled bison than bovine descriptions, mentioning hair and horns as the predominant feature. My experience using the bison in ceremony has connected with the group soul—and through ceremony, to the individuals of my herd in ways that are not heard of in modern agriculture. This is part of why I have brought them into use as the basis of the preps used in biodynamics for North America. Though their physical awareness can be threatening to anyone who isn't part of their lives, the thunder of a herd as well as the blowing of their breath into the air is an elemental churning of the earth's upper layers; their digestion and manure makes the circle complete, joining earth to animal nature.

Chapter Three
Animal Consciousness and the Individual

THE ANIMAL GROUP SOUL is carried individually and therefore we humans have not recognized it as being different from the way we experience our individual souls with respect to all humanity. The animals have a connection through group soul with each other that can be recognized by some people who are empathic, but more commonly we tend to think of them anthropomorphically—as individualized pets or units of production to be used on our terms. True, as individuals, they each have a part to play in the group—like bees who upon hatching, become house workers and grow through the hive to be harvesters, or are specialized into queens or drones. But insects have a more physical connection to group soul, as in the butterfly migrations are like a bird's migration instinct, which is also communicated through the group soul. We marvel at the flocks of birds flying in unison, like schools of fish, but it is really group soul that is individuated and acting as one.

When we recognize that all are linked spiritually through the group soul, we do what native peoples all around the world do when they depend on animals for their own survival. When you need something, and you are in a situation that doesn't provide access, there is a spiritual connection that people can use to access the animals' group soul. This ceremony is much like Steiner's approach with the biodynamic preps, a way to give the people access to the spiritual element in the ways of plants and animals (via prep materials) that will attract them to our farms for agricultural use.

This is an aspect of spirituality that Steiner doesn't specify, that the preps are a physical offering to spiritual nature. It is one of the first things I saw, coming from a Lakota background. The genius of what is in biodynamics connects directly to what I was practicing in the native ceremonies, making a physical presentation to spiritual nature. That is important in spiritual terms, as spiritual nature has no "reality" in this physical world, being timeless, voiceless, and bodiless—not defined in any way that our minds have been able to identify. Bringing that aspect of spirituality back into human experience (something we have been consciously ignorant of for a long time) takes faith in a concept that is in modern terms called religion, but is far more basic than that. The fact that native people have been chastised for "practicing a religion" that did not fit conventional ideas of spirituality is what brought me from my native studies of these connections to biodynamic farming.

Native people had a way of life that was intimately connected to the spiritual nature that Steiner regularly addressed. For me Steiner's Agricultural Course created an interesting path through religious icons (hierarchies of angelic beings) to the truth of native connection to the group soul of animals. I saw a link to the theory of Lakota people carrying the wisdom of the "lost tribes of Israel," which is now lost to most human populations on earth through the persecutions of various religions. Now this isn't exactly what is found in the individual animals, but it does mirror the fact that we now relate to our animals more as individuals than as the group within which they operate.

These individuals seem to us to be able to survive on their own, but what we don't see is that they operate as a group from birth—going through a metamorphosis of changes not as dramatic as a caterpillar to butterfly, but just as meaningful when the change from baby to adult is viewed from a spiritual perspective in relation to group soul. The parental instinct to protect its young is universal, as the young are certainly at risk and in nature that risk feeds the chain of learning. The young ones have an innocence and vulnerability that puts them at such risk of mortal dangers that they must learn the ways of life (and death) from the start. From the transition of umbilical blood, to breathing and eating, using the senses, the physical elements are a state of grace from the first day. It is during the bonding with the mother that the group soul connection is also developed. The whole group accepts this new life force into the inner circle of its being; many of the herd animals will form a physical circle around the newborn and that ring of protection helps protect it from the outer elements and predators.

I use my bison as an example of the family dynamic in that they resemble the old tribal hierarchy of a matriarchal society. It is one that values the youth and gives them experience and opportunity as they mature to adult status where they become family again, as breeding adults. The males have dominant status earned through size and agility within the family structure, but they will remain as sub-adults under a dominant older bull like warriors under a chief. However, it is interesting that the older females are the actual decision makers. Any female that calves is welcomed into this matriarchy. Males that are able to breed are also given a place, but that is usually reserved for the dominant bull who protects that right vigorously. That is the physical story of maturity and progression of the individual, while spiritually they are one family entity that has the generations at heart. The grass, sun, moon, and stars are also part of them through connection to spirits. The elementals of earth, air, fire, and water are some of the ways the individual connects to all things. The timeless sense of spirit that flows in all things is a thread of consciousness to which even humans can connect. It is an ancient way of human life that has mostly been abandoned, due to the influences of modern technology, which is referred to in Steiner's works as "Ahrimanic influence," something we should be very aware of and avoid exposure to if possible.

It is the stirring of the biodynamic preps, inspiring and enlivening by our hands the spiritual elements of life, that helps us to take responsible action in ending the individual life of an animal. My animals and I recognize the loss of the individual, and the emotional suffering at the physical level for the group as well as for the individual during the giving of life and death. When blood is spilled, the earth drinks it spiritually; it is like water germinating the spiritual nature of what we do in farming. Pouring water on the earth is done to keep the life forces flowing in the land. As my blood flows when I work on the land, so does the life blood in my thoughts and prayers, as I call out to the generational lines of ancestors to witness, and as the animal families recognize that this is not the end but a continuum of cycles. We recognize this transformation of matter into spirit, as spirit is transformed into matter. I share this richness of life in the sacrifice of the individual to the spiritual. Without the pain and suffering of birth, we wouldn't have the same transformation at death. Minimizing that pain and suffering is not the point, it is celebrating the transformation with responsibility, in honor, and with respect for the gift of life and death.

Chapter Four
The Spiritual Sheath of Animals

I LOOK BEHIND appearances and see the truth of what is behind having animals on a farm. It is a holistic view of what influences other kinds of animals also carry. Deer bring a unique feeling onto a farm. Like goats, deer have a way of seeing the world that gives them a special awareness of what they perceive. It may be that wind spirits, other elemental beings, and fairies interact with the physical presence of deer. The fire of their own egos takes a concentration of our will to control in the wild; it is what keeps the deer wild. This is also especially true of goats, even though they are domesticated animals.

The animal contribution is integral to the spiritual sheath that holds life forces on the farm. The grazing animal covers the farm physically in search of food, their harvest of plants is pruning the growth, stimulating regrowth—physically harvesting, digesting and returning microbes—as well as spiritually returning the enlivened essences to the land. In European spirituality, the gnomes are very anxious to inhabit this manure as it comes from the animals, for it is an enlivening substance for them to incarnate into, because of what it can become.

All the grazing animals need a family to reproduce: bulls,

cows, calves, or rams, ewes, and lambs. The choices a farmer makes in selecting livestock are based on his lifestyle and the land's capacity to support it consciously, while the biodynamic farmer's subconscious/superconscious mind is also aware of spiritual needs. These lead the mind in selecting what kind of farming to do. The land itself tells us what it is capable of doing.

Our animals lead a sequence of annual events based on the plants, controlled by weather, that we contribute to as little as possible. The decision of what animals we have is a crucial decision about what kind of life we share. The more we interfere with the natural lives of farm animals, the more we take away from their healthy instincts. We all learn more or less about our conscious relationship with animals through interaction with pets, and we recognize the difference in species, as in cats and dogs, birds and fish being such different types of awareness than our own.

In biodynamic farming, the preps' sheaths are animal energies that relate to the different species as well as to the organ that the sheath represents cosmically as well as spiritually. Specifically, with the 502 (horned stag bladder with yarrow blossoms), I was quite impressed with the specifics of its use and learning the "why" of the prep made me aware of the archetype that Steiner was describing in a stag. Its exterior awareness as well as the male aspect of its warrior energy and the specifics of using the bladder come from a long line of traditional people understanding the value in a receptacle connected to storage. The stag's kidneys and the character of its urine connect to the cosmos. The addition of plant morphology works, as yarrow has the same archetypal traits as a male deer, plus its own healing powers, which help to create the transformation into compost. And when that compost is combined with the larger sphere of our compost pile with the other preps, it shows the genius of Steiner's thoughts and actions in bringing us the preps as a spiritual way to affect human action.

When I first tried making the 500 Prep with different horns from male sheep and goats, I found that it didn't work. Though the material looked the same, when viewed from the spiritual perspective, it held no energy. As the animal archetype of sheep and goat is expressive, it cannot hold what is formed; it is not impressive like the archetype of cows—and it turns out, bison also. When I made my first bison horn manure in 2003, I was excited to discover I had an exceptional result.

Horses bring a huge amount of ego onto a farm. Because of that, they attract a completely different kind of spirit, one that takes a lot of physical energy to support; they have an immature spiritual sense of being—and are looking for the completeness we get to provide as caretakers. The faeries like the horses, but the gnomes are always trying to be the ones who are there for them. Horses' manure shows their immature characteristics, which need much more development to be a tool providing for all the needs of the plants.

Chickens carry a very powerful energy of selflessness, giving energy away in nervous bursts that attract the salamander/

fire spirit. The manure from chickens is also full of this same energy, a high-powered fertility they willingly share as they also shed eggs in a fertile blessing of the earth.

I do see why Steiner likes the cow; its archetype of earth/stone, and people/gnomes locks into the ground the energies of the cosmos. Their inward-looking mind is perfect for chewing cud and the contemplation of the mother-sense, digesting plants for the spirits, which is when the manure completes their part in giving. The bison share that earth-sense with cows, giving from a group soul, a completing part of themselves individually—but the process attracts the gnomes to the energy of the manure, which is not like that of the other animals. Bison have an intensity toward their group soul which is much more powerful than the cow or other bovine, likely related to the nature of living in the open rather than domestication. The contention is that bison are wild, but their nature has a spiritual character that is the same as the bovine. In North America, the buffalo's long history with native people reflects a direct spiritual connection, expressing ways that predate the agriculture of Europe. The native model avoids the European domestication issues of caring for livestock while losing the spiritual connection with the group soul of the animals.

Each animal is a physical sheath for this group soul energy, distributing its unique signature of *I am here*, just like the farmer whose energy expended on the farm gives a special flavor to the energy of how that farm feels. Straightening up or building a storage area, fencing repairs, and painting walls all take energy that is marked by the individual doing it. Hired hands contribute more than they think when simply picking beans and washing lettuce. The year's cycle is as fruitful in spiritual influences from our actions as the cycles of plants and animals are to the ground they grow in. Animals on the farm leave traces of influence, weaving these trails together as a recording of songs makes an album of music. I always say to anyone helping me to make sure to sing while you work—it puts your mind into the state of giving rather than thinking; then the hands can do the work of Spirit. Singing while harvesting especially revitalizes the plants, for you are giving them something that really stimulates the life forces even as you harvest the products of those forces, it is gratitude that gives as much as you take.

Each animal archetype distributes a different energy and that energy impacts every square foot of the farm; the farm is "treated" by the animals that are kept on the farm. The different sheaths that the animals themselves are, like the prep sheaths, provide different qualities to the material they hold, and that quality is individual too, just as each animal exists individually, as well as within the group soul presence of its archetype. This creates the "terroir" of each farm as an individual expression of the animals, and the farmer as a collaboration of energy that works with the plants to provide for the health and welfare of much more than the products that flow out into the world.

CHAPTER FIVE
Buffalo, Biodynamics, and Native Ceremony

THIS CHAPTER is about my vision for an agricultural future for the American continent, in which organic farming would be the new form of commercial agriculture, biodynamics would be practiced with bison instead of cows, and the preps would be made with bison who have experienced native ceremony as their life force is used for the biodynamic process. That kind of agriculture is needed to reduce the suffering of modern society, all the ills that plague the 21st century that stem from nutritional deficiencies in the food and the toxic ways it is produced.

This change would begin with the remarkable buffalo of North America—a story that starts with the development of a native spiritual connection with all things, a mystery of our own history that has been sadly overlooked. The bison's ancient counterparts were much like the horse. In America, the ancient forms were here and went extinct, but unlike the horse, buffalo showed up again in much the same form to repopulate the grasslands. Here is an animal that is sustainably adapted to this environment and occupies an important niche, but we have replaced it with cows—mostly due to their ease of domestication. But this also occurred in the United States as a government policy for managing native people, by reducing their food supply and making them dependent on the American political process instead. The main ingredient in restoring bison to a healthy form of agriculture is honoring the return of native spiritual practice. The bison carry that honor in their way of life and can restore honor to the native people who have been long abused. This would also restore the nature of the land to balance, instead of being overly exploited for its natural resources by plowing and mining. The minerals and oil, even the aquifers underground are being contaminated by our exploitative ways. It would be universally beneficial to use the resources of this continent wisely and with respect to the spiritual sheath that contains them.

My particular place in realizing all of this is my personal history of discovering native ways, and through that, grasping the aspect within agriculture that is spiritual, then finding the cross-reference of biodynamic practice. This combination could bring honor back to the native way of life and respect for the original people who practiced it.

In practice, in physical reality the animal does not know the specifics of what I am asking when I approach it. The action I take is like the biodynamic spraying of the preps, a physical representation of spiritual powers. The germination of this energy is the difference in producing the gift of what the animal/group soul offers and the sadness of what the archetype/group soul loses. The individual animal knows something is happening, and the group/herd also reacts to the ceremony—usually ostracizing the individual, who often will hang out or show up to the kill area I have chosen. In many cases though, the one who is chosen by the herd is the young adult who doesn't fit into the herd/flock and is often found on the outer edges of the family unit—vulnerable to this process of elimination. On the day of the kill, it is the ease of finding the right time to place the blade that always amazes me. My conscious action is seeing the heart and putting the blade into it—though I do miss fairly often. Usually I sing to the herd and the individual while waiting for the moment to come. The most amazing thing about this "style of hunting" is that I never practice: just one arrow, one shot, one spear throw, one buffalo, year in and year out. The fact that it is rarely over 20 feet away has a lot to do with it. My motto is *if you can't hit it with a rock, don't try to kill it*. The animal is there for this reason, but not to show how far out can I hit it as a target. From my point of view, that is the animal's consciousness of the process. There is a lot involved when an animal dies in this manner; it is about their consciousness in connection with the group soul and the separating of their spirit from that group soul, when the body falls to the earth. That is why it is necessary to keep the animal conscious at this time rather than yielding to the typical human concern of separating its consciousness from its body before killing it, to prevent individual "suffering." The pain and suffering of an animal dying is as normal to life as being born into it is; I have realized that when we tamper with it, we take away the connection of the group soul to the individual when we "humanely" kill animals. This is what we have been doing for the past hundred years by knocking them unconscious before taking their lives.

My process of learning about these things started when I was introduced to the traditional sweat lodge ceremony by a full-blood elder, Wallace Black Elk in the 1980s. I realized that what he said about ceremony was very true for me—that "these ways are for the health of the two-legged ones, the people." These ancient ways that so many of today's native and non-native people have rediscovered give us as an intimate connection to the world around us, a vantage point within our inner selves. I now carry a sacred pipe myself and often use it in the traditional native ways.

The buffalo ceremony uses tobacco prayer ties and songs in a four-day event, kicked off by a sweat lodge and ended when I smoke the pipe on the morning of the kill. At that time I use my blade without fear or anger to reach the buffalo that is ready to accept it. This leads to a quiet death where the animal is with his relatives. They aren't necessarily quiet when he goes down, and I am singing with the prayer ties ready as well as an eagle fan to quiet them, taking that responsibility for this ceremony. It constantly amazes me that I can walk into my herd of buffalo with fresh blood on me, and they make a place for me to place the prayer ties on the horns of the one I just killed. And within a few minutes, they mostly say goodbye and move off. Usually the big bull says farewell, as it is his responsibility to protect his family, and I have to let him know I am caring for the one killed now. I do this with the fan to signal "taking" this responsibility from him and return him to his living relatives. I have used this ceremony on over 80 buffalo, and it continues to work. When people have brought fear or anger to the ceremony and put me at risk, then I have had to use a gun to kill buffalo instead. This has happened fifteen times over the years. It also happened due to scheduling, when I could not complete the ceremony and had to do it within a certain timeframe.

It is obvious when it does happen, that something is very different when killing an animal with a gun, compared to using a blade. The gun breaks many connections with the shot—both the sound and the shock of it do more damage than we can see. I have been present (with my spear) at Sundance ceremonies where someone was giving a buffalo for the ceremony; the people often pray for the animal and even for the gun used, but when the animal dies, there is sadness and grief, which is the opposite reaction to my ceremony, where there is a celebration of the gift of life, honor, and respect. The difference is quite dramatic.

The huge number of stories about how this ceremony is so closely tied to connecting with bison has shown me that it has been used for a long time. It is possible to bring the bison into contact with people when needed. Because of what has happened to me, I know beyond doubt that bison understand what is transmitted through the ceremony. When I have been thousands of miles away in a sweat lodge, they have consistently responded to the requests I made of them in the lodge, such as going back inside their pasture fences if they had gotten out. I have also found that I can establish a relationship that is beneficial to the farm through ceremony with other animals in the wild. Of course it is also easy to mess up this delicate ceremony with human carelessness if we neglect to eliminate mind-altering substances and are not being absolutely clear in our communication with spiritual connections. Anyone involved in the process may bring in these problems, and doing that will "muddy the waters," so that spirits being asked to help in a confusing way, will do something not clearly defined and sometimes the opposite of what is being called for. As Spirit is in a timeless and emotional state of consciousness not embodied in a limited life form, alcohol and drugs are definite ways to alter our consciousness and consistently interfere with connection through these ceremonies. There are other ceremonies which use certain drugs to induce a spiritual sense, but they have nothing to do with this ceremony, which needs a clear and focused consciousness.

When dealing with the physical presence of bison, it is necessary to have all your physical reactions intact, for as slow seeming as they are, they also have very strong and quick reactions to any perceived threat. If they are ever trapped, they will throw themselves into instant rebellion against any threat, which makes the saying "you can make a buffalo do whatever it wants to do" a very true thing. Managing bison is all about understanding what they want to do and letting them do it, in a way that also accomplishes what *you* need to manage them. This is why these ceremonies are all about learning to understand the bison and letting them know what you need and that what they need can be done at the same time.

The reason for using a blade has been made clear when all those involved are participating in a celebration of the gift of life rather than the loss of life when a gun is used. The use of a gun represents a tearing of the conscious life forces, as well as the literal loss of consciousness with brain trauma, in comparison to the relatively gentle loss of consciousness when a blade is used, and life ceases more slowly through blood loss, internally in most cases.

Buffalo Day

Sweat lodge four days earlier, with prayer ties on the altar. Songs sung, sacred pipe smoked, sunrise coming. Ready, my heart in butterfly stage.

Arrows and bow, spear as back-up, frosty air lightens into daylight. Animals breathing clouds of steamy breath, quiet in the air. Eagle feathers rustle in my hands. Prayer ties bright in the drum frame. Beater taps impatiently, but it is time.

Feeding as usual, animals come in, my prayers are with the one chosen. A young bull, standing quietly, apart, slightly to the side, there is his heart.

Pull the bow, aim and breathe into release, thunk, into his side, behind the front leg. He jumps and staggers off a hundred feet, standing, breathing heavy, blood coming out where the arrow protrudes. Excited family pushing and shoving him to stay upright. I am singing the piercing song as he staggers and goes down. Big bull hooks him, snorting and pulling him up.

I step in with the drum, singing, holding the eagle fan and the prayer ties. Blood on the noses of his relatives, sniffing him. I take responsibility and wave them back. Big bull snorts and paws, but pulls back when the older girls realize what I am doing, they have seen this and it is okay. He backs off and I step up, fanning off the last breaths, standing clear if he has a kick left in him. But he is quiet now. Blue eyes staring at nothing. Bright blood on the grass. Buffalo family standing as I put the prayer ties on his horns. I leave, and they make sure he isn't there any more. When I come up with the tractor, the big bull is the only one waiting for me, and he is just protecting his territory.

Again I take responsibility with my eagle fan. The power an eagle has is also a connection I learned from the native people; I honor this bird and what it represents in spiritual ways. The responsibility and care that is inherent in the feathers is about the lifestyle of living with the earth and being a warrior for the earth. Each feather represents an act of service to the people for the earth, acts of bravery, courage, endurance, and of the many things that one can do for others. It is also about the "giveaway" and "the gifting of self." It supports the "no fear or anger" state that I have to hold when facing my herd bull, at this time. When he turns away, I rest the fan on the tractor as I hook up chains to the slain animal's legs and back up the tractor. The bull stops at the gate as I pull the carcass out of the pasture.

To make this action into something contemporary, we need to relate these ancient mysteries to our modern day practices, which is where the biodynamic work applies. Following the 1924 Agriculture Course's basic premise, we have developed a relationship with the natural and spiritual practice of making food into a substance that is a fully enlivened part of life again. We need food capable of carrying out its part in sustaining life, rather than being "fast food" with poor nutritional value, which must be supplemented by taking more pills, made from some truly scary sources.

From my viewpoint, based on native traditions on this land, it is easy and natural to choose making the preps from animals representative of the archetype I identify as being "of earth" here, the bison. Linking native practices and connections through ceremony to the biodynamic building of the farm organism is a natural way to bring the parts together as a holistic expression of the native concern for "all our relatives." I look at the preps as prayer offerings to the spirits of the land, much as the native prayer ties of tobacco in colored cloth are presented to the spirits as an offering. Tobacco is such a powerful herb for holding the essence of our prayer to the invited spirits in ceremony to hear and choose to answer. The biodynamic preps are so specific in their makeup that our individual prayers are a backdrop to what Steiner has put forth as a powerful invitation to do the work on the farm from a spiritual perspective, for the purpose of restoring the "farm organism" to its highest state. These powerful "essences" of form, as in the oak bark, in the brain space, in the yarrow in the stag's bladder—are powerfully reinforced as spiritual enticements to do specific work in the compost. As "homeopathic" remedies for the landscape, the horn manure and crystal preps are also specifically applied at dawn or dusk to enhance the intention of enticing the spiritual enhancement of the application. Each of these preps is also related to the planetary bodies, as follows: the horn manure (500) is related to the earth energy, the horn crystal (501) is for the sun energy. The stag bladder with yarrow (502) is Venus, the dandelion and mesentery (503) is Jupiter, chamomile with intestine (504) is Mercury. One of my favorites is the stinging nettle (505), which is a Mars influence, though it represents the heart of the compost preps in the pile. (It is a Mars influence in the planetary lineup.) I have been using the pericardium to wrap this prep, though it typically is not wrapped in a sheath. The oak bark in the brain cavity (506) is buried in the watery element of flowing water in the earth, which relates to the moon, and we also have valerian flowers made into juice and fermented as 507, representing Saturn. I'd like to especially thank Heike-Marie Eubanks of the Oregon group for remembering these.

These are the compost preps used for bringing the earth into balance with the cosmos, put into the earth for transformation. They start the process of bio-remediation by stabilizing their own changing status as individual elements of creation by being wrapped in the membranes which hold certain energies in animals' bodies for massive transformation.

The timing of putting the preps into the ground for their own stabilization of energies relates to their cosmic relationship as well as to their animal sheaths' transformative representation. The physical structure of the sheath is indicative of the energy it transmits from the flower energy of that flower's life force that remains as an imprint of archetype. The echo of the flower's other common uses relates to why this combination of sheath and flower is used. The digestive use of chamomile indicates that we should use the sheath of the intestine to hold that energy in the compost, even though it is more obscure in relating to the planet Mercury; that involves understanding the workings of astrology, which is another field of study that is not well-interpreted today. This is a kind of alchemy that is not known in science and seems to put the whole idea of biodynamic agriculture into a class by itself, which it is. This is a spiritual science that uses intuition and clairvoyance as very real ways of relating to the elemental balance of life forces. Understanding them in ways not described by the rational mind makes it hard for the reality of their significance to be accepted. But that reality is used all over the world with great efficacy. It works in the mysterious ways of spirituality that have been practiced by people more connected to the earth than is modern mankind. It is easy to recognize the spiritual familiarity in Steiner's work when you have familiarity with the spiritual usage of ceremonies yourself, even from a culture as distant from modern life as the Lakota. This goes to show that there is a universal spirituality which applies equally anywhere in the world, and that is also what I see as the appeal in biodynamic agriculture. It applies universally wherever it is practiced too.

CHAPTER SIX
Birth and Death in the Group Soul

IT IS THE STIRRING OF THE PREPS by our hands that inspires and enlivens the spiritual elements of life when we take responsible action in ending the individual life of an animal. My animals and I recognize the loss of the individuality and the emotional suffering at the physical level when I am taking this life. You can see the way it breaks family ties and destroys shared memories whenever life is ending.

When blood is spilled, the earth drinks it spiritually; it is like water germinating the spiritual seeds, similar to what we do in farming. Pouring water on the earth keeps the life forces flowing in the land. As our blood flows when we work on the land, so does my life blood flow through my thoughts and prayers. I call out to the generational lines of ancestors (group soul) to witness as the animal families recognize that this is not the end, but a continuum of cycles. I have seen the difference in recognition of this continuum by group soul when my flock of sheep bed down on the blood of the one I killed earlier with no concern for the outcome, while when I shear a flock that has had "the butcher" harvesting, they are frantic to get out of the way when I walk among them and are traumatized when I handle them.

That we may recognize this transformation of matter into

spirit, as spirit was formerly transformed into matter at birth, the ceremony helps us observe what this does to the group soul. In honoring the one we take life from, it is given a place to transform from the physical into the spiritual form which we each carry inside us, an incarnate spirit that is described as bonded to our soul. That spirit is released at the freeing of the soul from the physical life we live. In the simplest of terms, it is the incarnate elemental being that is within everything that is released whenever the physical form is changed into something else. Even at the level of the senses of taste, when we release the flavor in an apple, it is the elemental released to reincarnate into another being. I share this richness of life in the sacrifice of the individual to the spiritual. Without the pain and suffering of birth we wouldn't have the same transformation at death. Minimizing that pain and suffering is not the point, it is celebrating the transformation with responsibility, honor and respect for the gift of life and death.

On our farms healthy young animals are usually the ones who are sacrificed. To harvest these animals is to create a space for the cycle to continue. In conventional farming, that usually involves shipping them to slaughter at an abattoir certified and approved by the "humane" methods of mass killing. The alternatives are to have a farm butcher truck come to the farm to take the individuals. But it has to be by private sale, as the meat is not inspected by the authorities, who have nothing to do with how it was raised or how it is to be killed. This is where the consumer can have a say in how it is done.

Both Kosher and Halal have basically the same principle for taking life and are based on this spiritual connection that is vital to the return of the individual's spiritual connection to group soul. This is something we as humans haven't been able to acknowledge in our "humane" way of thinking, as we are typically single-souled entities walking around individually. We operate singly and think in that form, not related to each other and only distantly to the group soul consciousness of these animals.

The animals' conscious lives are the fruition of the annual cycle of the year. The animal families we take care of—the goats, sheep, pig, chickens, cattle, and so forth—are a continuum of life forces that are entwined in the process of sunlight and green coming from the earth, connecting consciousness from the cosmos through the work that they do on the farm. Their work of digestion, procreation, and the intimate play of health are life forces pulsing from the stars and planets, through the seasons of Earth's turning in the cycles of our year's weather, patterning the annual growth of plants. The interplay of insects and animals, with the unseen spirit of nature, conspire to create the fabric of life. That spirit is a counterpart of the elemental world of earth, wind, water, and fire—it is *there* that we have lost contact with the consciousness of animals.

The lives lived in contact with all these relationships with nature have another relationship within themselves, that is the group soul, an interconnection of the individual (as close to individual spirit as we are) to the archetypal being of their kind. This group soul entity in the individual connects through the life force of consciousness to the group soul family and archetype. That connection gives each individual in its development the sense of "being" that humans refer to as "instinct." It is like a bird's migration and animals' seasonal reactions to the elements. This group soul mentality does not include humans, as we have mostly grown individualized in our relationship to the world, becoming a "higher life form" in terms of responsibility for ourselves. We have also separated ourselves from the spirit of nature with regards to the consciousness of animals. It's not that we don't have those feelings of connection, but we lack a *consciousness* of them. This allows us the development of mind and body with forms of thought that plants and animals have no cognition of, making our technology as advanced as it has become, but these forms of thought have usurped our senses on many levels. The dangers of our unconsciousness have not been evident as the spiritual world is timeless. That fact has given us the "time" to realize our folly before the results of our lapse in recognition of the timeless spirit of nature proves completely fatal. That sense can still be recovered in a way to save humanity from itself. The senselessness of which I speak has become obvious in the evidence of planetary changes that point to our lack of awareness in dealing with these problems. Recovering our senses is inherently a part of the process of finding the solutions that we need. That recovery could help us accomplish what we need to in order to survive, at exactly the same rate of discovery that new technologies show today.

An important way for us to rediscover ourselves is grasping what is really involved in the death of animals. That death is usually at the hands of a species that does not have a group soul mentality: humanity. We address the animal, concerned with its physical suffering from that individual death, not realizing the non-physical connection that animal has with group soul. Immersed in own our humanity, we have addressed this with techniques to take a life without much suffering that are labeled "humane." These methods take consciousness first, by separation from mental process through shock to the brain, achieved by anything from a bullet, a contained bolt that goes into and out of the brain, a hammer, or electricity that short circuits the brain activity—all resulting in a so-called "humane" aspect of loss of consciousness. That is before the actual taking of the life force through letting of blood, by cutting the throat arteries. These are the "approved" methods of killing in modern abattoirs that are considered "humane" by the science of modern medicine and sanctioned around the world.

The modern medical method of putting an animal down is by lethal injection, where an anesthetic is administered directly to the brain via the carotid arteries, causing immediate collapse of consciousness and brain activities. The kindness is more to the owners who don't want to see the animal suffering, which is all well and good in the manner of humans. In a world of pain and suffering, we have extended our own lives beyond reasonable expectations with all the medical knowledge of modern science. And, in our extended suffering from diseases such as cancer and other trauma to the body, there is much more suffering over time in our human condition than most animals have in their entire lives. Healthy animals living in natural settings have enjoyed so much of their lives in the embrace of the group soul. Compare that to our lives lived in servitude to jobs in urban settings of artificial comfort and security, where we have created a false sense of reality that serves us rather than us serving it. The manner of death that comes in nature is usually from predators or accidents to the younger members of a group soul who are in the process of gaining experience and end up making fatal mistakes. These are, in human terms, the "formative years"—when parents and community are the teachers. Like most communities of group soul, we humans gather our mental resources for teaching, as group soul interactions of physical reality do for the animals. But the link of consciousness is vital to group soul—especially at the time of death— so the dying animal can experience the "instinctive" process of its spirit separating from the group soul. To arrange matters otherwise causes a weakening of that instinct in our animals. This removal of the group soul connection is something we have been doing to our animals for a century now. This is not something you can grasp in a year to year progression. But the erosion of the spirit of life in all things is what is being addressed within the holistic system of biodynamic agriculture and indigenous culture the world over. It has only come to my attention through the killing of my bison with the age-old technique of native ceremony and using a blade to take the life force. That is how I learned that the process of "humane" killing is taking away a part of the animals' connection to the group soul, a thing you never are aware of otherwise. There is a different wisdom that was held by our ancestors in dealing with agriculture, and we can access it when we deal with our animals hand-to-hand. That has been lost with our technology, but when the quaint "old ways" are used, there is a bond with animals that is always available and instantly activated, giving us that sense of connection. For example, in the hand-milking of a cow, there is a connection to the giving of life force that comes through group soul, one that we as humans don't usually recognize we are so individualized spiritually. That connection through touching animals can restore our sense of vitality as individual representatives of group soul. It really is as simple as touching our animals; even the pets of our households offer this sense. It is easy to see in the difference in cats' and dogs' group souls, as well as in their "giving" of themselves in the sense of life forces. The recognition of group sense is found in all animals. We have only lost it temporarily.

Becoming aware of group soul in the way we handle our animals is an important way to bring it back to our consciousness. My proposal is to certify the on-farm butcher services as "conscious killing" as opposed to "humane killing," a change which will need the support of the consumers. It would include attention to the life forces involved and would be along the lines of Kosher and Halal certification and labeling. It would involve a much more conscious technique of taking life, recognizing the giving of life by the individual being killed. It would include awareness of the severing of the conscious link of the individual to the group when the "butcher" kills it by the use of a blade, either by stabbing through the lungs (or heart) of a large animal or in smaller ones by cutting the throat, but not removing the head or severing the windpipe. This permits a period of time for the crossing of physical and spiritual boundaries that is like a birthing but is its opposite, a death process that all creatures should be aware of consciously.

CHAPTER SEVEN
Blood and Guts

THE INVITATION OF SPIRITUAL INFLUENCES does not need to be understood in the mental sense, indeed it is not understood, as it cannot be defined by rational minds. *It is the heart influence, like love, that we humans can feel that is like the spiritual energy in biodynamics.* There we have the vessel of life, a lifeless body, the intricacies of heart, lung, muscle, bone, hair, hide, and the mysteries of endocrinology and nervous systems. These are a gift from spirit to heal the ailments of mankind, taken from one life to heal another, the life we still have. That is the prayer behind all food and the invocation that some people continue to make when sitting down to eat. An acknowledgment of what is being offered is all that is needed, just a nod of the head is the physical reality that Spirit needs to be part of our reality.

With the bison harvest, after I have invoked the spiritual, the blade is applied, the animal dies consciously, bleeding out, losing consciousness, the physical body and brain disassociating, and finally the heart stopping. The lungs exhaust, it all stops, the final spasms are like an attempt at physical escape, the individual's last act. Some animals are quiet, some kick for a couple minutes—an individual expression of archetypal fighting or running away. Much like the first breaths and spasms of life, there is a wonderful fulfillment at both the birth and death experience, when it is honored with respect and understanding by offering song and prayer ties on the horns. After we have addressed the consciousness, we have the wonderful medicine of Spirit. The physical form of the organism, the exterior hair and the silica process of exterior expression in the horns and hooves, where the animal is touched astrally by the spheres, is the means of making cosmic connection between the stars and Earth.

Now it is becoming the flesh of the earth, transformed into buttons, spoons, combs, yarn, padding, rope, and so on, the skin and membranes of muscles become thread and fiber for sewing and leather. Muscles turn into the meat of our meals—the protein—and bones become tools, containing the sweet marrow. Organs contained inwardly are the treasures of deepest thought—including the separated consciousness of the metabolic heart and lungs, the endocrinology of the digestion, guts filled with the contents of the world. This is the complicated cosmology of the universe contained within an animal.

This is all put to practical use after consciousness has left it. The nutrient rich organs hold value medicinally as well as spiritually and cosmically. When Steiner wrote, these organs were still being discovered medically, but he was already discussing them in his work. The lymphatic system is extensively tied together throughout the internal structure of the body, linked through the nodes at every junction of movement. This enables it to be "pumped" by the actions of the muscles, like a heart that is used when it is needed, moving it where the muscles are being used. The system carries lymph, but what is the description of lymph? A fluid that travels between the cells throughout the body, carrying the things that keep us healthy, linked from bone marrow to the thymus, kidneys to the heart. The lymph touches us internally, making seldom heard of links like the appendix to the tonsils. It carries the T-cells and white blood cells that are the forefront of today's medical "fight" to keep us healthy. It has always been there working, almost spiritually transporting bone marrow T-cells to the blood, where they are transported to where they are needed again. The lymph system goes everywhere internally, throughout every space in a timely manner, doing things normally thought of as impossible. It brings substances locked in the center of our densest structure (bones) into our most fluid one, the bloodstream. It cleanses the body of our foulest detritus, taking it out through the spleen, pancreas, and less so, via the liver and kidneys. But it also brings the healthiest resources where they are needed. This system is a mysterious network that is a "hand of god" behind most of the healing influences at work in the intricate connections within all bodies. It is interesting that the most common form of modern medicine used to "heal" us is removing the affected organ, especially those linked to the lymphatic system: the tonsils, spleen, or appendix—as they are the first sign of a system being compromised and thus becoming a liability to the rest of the system. I would suggest looking into the healing of a diseased organ as a key to restoring health rather than removing it from the body, as we are not a machine with replaceable parts, which unfortunately many practitioners and patients are being led to believe.

Next, in this survey of the animal body, we look into the organs of thought, literally the brains, something that Steiner compares with the results of our digestion. After consciousness has left them, they do become a mass of enzymatic pus; I see it as having the contents needed for the practical use of tanning hides. Each brain contains the enzymes needed for the successful preservation of the skin for use as leather. Hmm, more food for thought?

There is also some modern healing being done by using the thymus and adrenals as cures for those with compromised immune systems. I would suspect that internal experts are already hard at work disassociating the parts of the lymph nodes themselves, in order to get at the importance of the structures as portals of inner being. Though the importance of the adrenals and thymus applies to compromised immune systems, they are not generally used in healthy individuals. The more gross organs are the lungs; their importance as food is not well recognized, as they are mostly a transformational organ to bring air safely into our being, exposing the blood to air on a microscopic level rather than injecting it as an embolus. Amazing as it is, the lungs don't relate well to nutrition; though they are edible, it is not the choice for many, except for invertebrates.

Then there is the heart, a much maligned organ that is really a muscle, holding so much significance in our minds as the one that transports us into the realms of love, but really it is just a reliable form of transport for blood, like a literal pump. Steiner refers to it as being more of a flow form that helps the blood move than a pump that pushes it, which has led many to discover the liquid forms that abound in nature to revitalize the fluids. This discovery also draws attention to the vortices and the channels that rivers naturally flow in. From tumbling creeks to the meandering of rivers, water has a constant purpose and carries much of the same functions as the blood—cleansing by using the vortices of nature, exposing the fluid to much more microbial activity than any other kind of stirring. Of course in the heart and body, the blood must also not coagulate, so the constant movement keeps it alive and the "pumps" assist in moving it smoothly like the vortices in water. The activity of muscles also helps the circulation by providing a pumping action like gravity in the fall of water in nature. All of this makes the heart a wonderful vehicle for being the center of feelings in a body.

The way we handle making animals into food for us matters. I have described the ceremonies I use for the actual kill. But I continue acting with reverence as I process the carcass and prepare the meat for sale. We address our physical health with much more than food. I let the bison's blood onto the ground (catching some for use) and get my tractor as the herd is also accepting the loss of the individual. I remove the carcass to my outside abattoir, raising the body by hind legs and start skinning by opening it tail to throat and across the legs. Once skinned with legs and head removed, I open the body cavity taking out the organs and guts, harvesting the prep sheaths and fat, then I let the body cool for a few hours up to overnight. Afterwards, I cut and wrap the meat for freezing until it is sold. I only sell locally person to person as a commitment to keeping the circle of life connected from the animal to the consumer as medicine for the people. I do not ship it or sell it through other handlers. This keeps the medicine intact from the animal consciousness to the human consciousness. Not many of the people who buy it realize the potential of what we do in biodynamics or in the process of making this way of life available, but we still do it because it is a way of life that provides all we need to survive as a species with the planet intact to share with future generations. I do it this way to offer health and help to the people, animals, and land.

II

Poetry
by Devon Strong

Buffalo and the Blade

I dreamed of hunting buffalo.
The old days and ways of preparation,
singing old songs,
sweat lodge,
and looking across the plains,
searching out the buffalo family.
The one that is calling to me,
I know it.
Seeing the black dots,
knowing them to be buffalo,
not boulders, deer, antelope or cows.
Those are buffalo,
I know it.

Walking up on them,
gently moving them towards the surround,
I smell them,
they are buffalo.
I follow them across the plains,
I see the one I carry my lance for.
He carries his heart there.

A Lakota Approach to Biodynamics

I sing the old ways,
to the heart of the buffalo nation,
the sacred hoop of health and help,
without fear or anger.
I look at this animal,
I smell him, I see his fur,
shaggy legs and bearded head, set with horns,
looking for a way out of the surround.
This time I will show him a new way.
I fill my pipe with courage, strength, endurance,
and not forgetting patience.
I smoke it up,
waiting for the place to open up,
accepting this
spear into his heart,
that place behind the leg,
down low, up front, right there!
Its point is made,
the blade is sharp,
a quiet energy picks me up and puts the spear where it goes.
Letting fly, it sticks him,
with a grunt he kicks it out, takes a few steps,
staggering with a strange quiver,
his heart is pierced,
a gush of blood comes from his wound and suddenly
he can't stand, losing balance he falls.
Lying quietly, last breaths fading,
a few kicks and there is a great

honoring in this gift of life.
Food and drums, hair and bones.
Seeing the soft blue eyes fading,
bright blood in the dust,
tobacco ties twined in the massive horns,
hooves lying quiet that beat a drum of earth,
this is not something lost, but found.

Honoring the way of life and death,
offering my hand to the buffalo,
I know it.

Now the sharp blades cut the hide,
opening the way to meat and bone,
and peeling the hide back exposes the insides,
tender parts of liver, kidney and
the great digestive tract of guts and
gossamer tracery of blood and fat, and the mesentery.
Lungs and heart, bloody with the freshness
of death, smell of meat, guts,
and hide, within the hint
of wild that is buffalo,
I know this.
There is a special time of sharing,
with all the beings I offer a gift of flesh,
to eat of this is to sustain life,
I know this.

The offering is from the buffalo,
but it is part of who I am,
I know this.

                                                              10/6/06

Coming Together
a prose poem

This is a strange time when people from anywhere in the world can come together and talk about the ways we share in living.

I was talking about the Mayans of Central America, and Morongo Ethiopians sharing similar fire ceremonies today on a hillside in northern California after a peyote ceremony from Mexico and a sweat lodge from the plains attended by northwest Haida elders.

Some question the integrity of "mixing" traditions, but this is an honoring of the integrity of the traditions through sharing. My recent experience of a traditional local sweat, Karuk style, was another sharing, as was a recent gathering youth and elders conference that was intertribal.

From Nez Perce to Klamath, Karuk, Lakota, and many others, from Mexico to Canada, all teaching the variety of traditions that are spiritual in nature and traditional in spirit. I am seeing and becoming part of a growing multicultural heritage for our children to experience as something more than a reality TV show. The challenge is to live in a way that is producing the sensual experience of life in the old ways, while living in touch with modern world concerns. This is, in fact, bringing a cutting edge way of developing "green living" into our local community.

This world I live in is part of two ways of life, native-traditional, and traditional-American.

I am both, and true to my own belief in living as Americans, who still have the freedom to live according to their stature in life, whether taking all they can or getting by the best they know how.

Still, there is a freedom that is lacking in many parts of the world that is shared in these gatherings.

Persecution makes headlines daily about atrocities against native people. But, whether they have been there generations or eons, they are still there, marking time, adopting patience through necessity and believing faithfully in spirit.

Traditional ways keep up with the times, as adaptable as the Bible and just as religious.

8/08

III
What Others Say

BISON AND THE SACRED HOOP OF LIFE
by Catherine Preus

LONG DARK PONY TAIL, muscular body, odor of sheep and the barnyard, perpetually smiling, weather-reddened round face, Devon Strong filled my Mount Shasta apartment with his presence. He handed me frozen buffalo shank in two small packets neatly labeled and wrapped in white paper, then proffered a third packet. "I brought you some London broil to try." As he handed me the little packets in an almost reverent manner, I felt a special thankfulness for the medicine, as Devon calls it, and the sacredness of this transaction, for I knew I was participating in what Devon calls the Sacred Hoop of Life.

After fifteen years of vegetarianism I find that my body does better with occasional small amounts of red meat, and I buy Devon's buffalo meat because of the care and spiritual awareness with which Devon raises and harvests the buffalo. The animals Devon raises on Four Eagles Farm in Montague fit into a twenty five year commitment to organic food production for local community.

Early on, Devon started the first Community Supported Agriculture (CSA) project in the Rogue Valley and Siskiyou regions; later, to help foster community and the cause of local foods, he was inspired to begin the Montague Farmer's Market; and now he is developing a Siskiyou Natural Farm Cooperative with a website for local farmers and ranchers to list his products and sell directly to consumers. These are just a few projects that mark the span of his commitment.

Today, Four Eagles Farm is the site of his work. Ten years ago Devon began raising bison, and three years ago, after years of renting land to raise crops and animals, Devon purchased 80 acres in Montague and moved the herd there.

"It was an empty flat piece of land with no water, and no one believed it was worth anything." He drilled and found abundant water, which probably increased the land's "worth," yet Devon stands apart from such notions. "I don't feel I own the land. This is for the generations," he says. "This is the answer to all my prayers."

He has planted fruit and nut trees and is planning strawbale/cob/adobe farm buildings. He owns sheep for triple use: wool, lamb and cheese making. In addition to farming and raising buffalo and sheep, Devon spends each spring traveling around the region in his red pickup shearing sheep for other farmers. "I can't get anyone else to do it," he says. "Nobody likes that work anymore." He also shoes horses, spins wool, makes felt blankets and is skilled in Aboriginal arts such as flint knapping. He is a Lakota singer and Pipe-Carrier.

Using prayer and Lakota ceremony along with ecological science, Devon operates his farm within the holistic philosophy of life the Lakota Sioux Indians call the "Sacred Hoop of Life." According to this way, all things that exist independently are also connected to one another in the One Spirit, called Wakan Tanka in the Lakota language.

Devon was raised on ranches in Nevada and taught a very different form of ranching from what he practices now. Though both methods are practical and results-oriented, they differ greatly in spiritual intent. As a child living in the Nevada desert, Devon began to wonder how Indians had survived in such a harsh environment without stores and farms to supply all their needs.

Years later he discovered the Lakota path through the teachings of Wallace Black Elk, a Lakota Elder based in southern Oregon at the time. "I just ate it up," Devon says. Now he is fully committed to keeping the tradition going. "It's here, it still works—I wish more people would do it!" says Devon. "The old traditions are all lying on the ground waiting to be picked up. I am trying to use the tools in ways they are meant for. "Other ranchers ask me why I run buffalo when it's so much easier to run cattle. I do it because I know how. Buffalo are completely different from cows." Devon's prayers and ceremonies and his assured quiet manner of handling the buffalo keep them feeling secure and calm. These other ranchers are amazed at the ease with which Devon can load a buffalo in a trailer and the method by which he retrieves strays, sometimes walking fifteen or twenty miles out on the land with a stick to bring them in closer.

Devon also slaughters, cuts, and packages the meat. Buffalo are so unpredictable and volatile and cause so many problems in meat packing plants that they are exempt from the

USDA meat inspection program—and thus Devon is able to harvest the meat himself and sell it. His method shows his philosophy in action.

Four days before the kill, Devon conducts a traditional Lakota sweat lodge ceremony. "I ask for a certain buffalo to come into the corral to be honored in this way: to create the sacred hoop, make way for the generations to come and be medicine for the people. All these different ways of asking them occur so that spiritually the road is opened up, the path is clean, everything is ready." Invariably the buffalo chosen shows up four days later in or near the corral. Devon joins the buffalo in the corral and kills it with a spear, while others hold a feather and beats the drum. If he spears the heart, the buffalo takes two steps and dies. If he spears the lung, the buffalo takes about ten minutes to die—all the while with singing, praying, and drumming in the background.

"It is absolutely peaceful," says Devon. "It's very quiet. You realize that the wind is blowing, the birds are singing, life is going on and life is passing over."

Instructions on how to kill the buffalo came to him in the sweat lodge. "It was as if somebody gave me a page with everything written out: Do it this way. Take four days, do the ceremony this way, sing these songs, make 150 prayer ties, and use a blade. The animal will be ready for you and it will happen like this." It also came to him that "a bullet to the brain is the ultimate violence; it's unnecessary. When somebody shoots a gun it's a spiritual shock to the animal, it's a psychic shock to the whole region—the ripping of the air with the bullet—there's a violence with a bullet that is completely lacking when you use a blade." The road out to Montague runs through a lonely, expansive landscape of ranches scattered in the hills and treeless plains. Over a rise in the road appears Four Eagles Farm's buffalo herd. Silhouetted against the sky, they look ancient and rugged, like a sepia-tinted Edward Curtis photograph, a relic of an ancient nomadic culture. I pull my car over, get out, and stand looking at them for a moment. Only the wind disturbs the silence of the high desert land, and the vast terrain dwarfs the herd. My thanks to them and to one man's vision and dedication carry across the high desert land and beyond.

CATHERINE PREUS *is an artist and writer living in Mt. Shasta—the town where "We're all here because we're not all there." She cares for elders, transcribes and loves to do in-depth interviews. She writes about the earth, sustainability, nutrition and explores our collective contemporary American grief at being a (hopefully recovering) rootless and transient culture. A previous version of the article appeared in a 2009 edition of* Edible Shasta Butte.

Letter from Jean-Michel Florin,
Co-director, School of Spiritual Science Section for Agriculture, Goetheanum; Dornach, Switzerland

Jean-Michel Florin
Sektionsleitung
Sektion für Landwirtschaft am Goetheanum
Hügelweg 59, Dornach

Dear Ms. Strong,

I facilitated Devon Strong's first workshop at the International Conference On Biodynamic Agriculture. It was for me a beautiful encounter, but since I don't know him well I can only write a few words about this session.

We learned of Devon from our working group about preparations. Uli Johannes König from the German Forschungsring told us about Devon and his way of doing horn manure preps with buffalo manure and horns, instead of with cow manure. So Devon was first invited to our conference to give a workshop about this. Last year we had a special conference about the theme of working with our animals in dignity. I proposed that we ask some people to give a lecture about this and organized a workshop about the killing of animals, which is a huge question in the society.

We invited two people to share different ideas about this topic: Stephane Cozon, a shepherd from the south of France and Devon Strong from California. We did not mean to imply that the old Indian way of killing chosen by Devon is the way for everybody today. No, the idea was to look at what is important, in order to offer the possibility to each farmer of preparing himself before killing animals with a modern "ritual" that each must find today.

For me it was interesting to see how Devon followed his own very individual way with biodynamics and the animals, and how he tried to bring together the traditions he learned from Indian people with the spirit of biodynamics. In this sense he was an example of how biodynamics and Anthroposophy shouldn't be a totally different thing that replaces what we already know, but should instead help us to deepen, and to widen our knowledge of our traditions and bring all of that to practical tasks like making preparations, killing animals, and so on. Devon was an example of this individual way of finding his own sovereignty as a farmer.

All the best,
Jean-Michel Florin

My Encounter with Devon Strong
by Uli Johannes König

IN JANUARY OF 2005 I was invited to a Biodynamic Conference in Sacramento, California. The subject was the preparatory work required for biodynamic development, and my job was to tell about the work of our research center here in Darmstadt, Germany. It was my first visit to the American West, and I was awed by the mighty natural forces one can encounter there. So I was filled with a marvelously mixed picture of gold mining history, Lakota culture, natural forces, and the "new" Biodynamic movement.

It was hardly obvious at first that among the participants was one man, Devon, who represented this mixture and went beyond it. Harald did call my attention to him as a person who had a unique understanding of the biodynamic project: he was a member by choice of an Indian nation, worked with buffalo rather than cows, prepared for the work with the organs of buffalo, practiced Indian ceremonies, and remained at the same time a dedicated Biodynamicist. He and his practices were only partially acknowledged by his colleagues, but that is the nature of American diversity.

When I saw Devon, I was somewhat amazed that such a calm, soft temperament could work with wild buffalo. I remained amazed and did not understand him, nor did I find the occasion to get to know him better then. What I remembered, essentially, was his gentle and friendly personality.

Then in 2010 there was an International Agricultural Conference in Dornach, Switzerland on the general theme of the Christian Impulse in Biodynamic Agriculture. The previous fall we who were delegates worked on planning the Conference and had found some echoes in Eastern culture: India with Hindu reference points, Egypt with Islamic background etc. I awoke to the fact that the West, with its tribal cultural components, was not represented. I said I know someone: "this guy in California with the buffalo!" We called Harald in Sacramento and asked him to invite Devon to make a contribution to the Conference. Devon accepted the invitation! Since the plans for the Conference had already been largely made and all Lecture openings were already filled, we asked him to offer a Workshop.

My encounter with Devon at the Conference in Dornach felt as though we were old friends! Though we had hardly noticed each other in Sacramento, we felt a deep connection at the conference. We had a number of conversations where we exchanged ideas, and I began to understand him better. I asked him, "How do you work with the animals? How do you create a relationship with them?" He said many things that seemed to me self-evident, though I had never heard them before. To prepare an animal for its death—yes, why not? But it seemed to me a miracle that the animal willingly comes to take part in its death ceremony.

We also exchanged thoughts on the way the basic Biodynamic relationship relates to Indian knowledge. When one really listens to the ideas expressed, one comes to understand the mutual connection, and beyond this, one recognizes the way contact between Indian culture and Biodynamic thought enriches both. Devon said to me that it was the greatest gift given to him in his life to discover Biodynamic agriculture. Being invited after Dornach to the Goetheanum exceeded his greatest dreams for himself. He was indeed overflowing with joy at that moment, as I could clearly feel.

Then there were other special moments that remain engraved in my memory. Again and again he described how the buffalo, compared to the cow, has a greater energy and how this matters for the nature of the American continent. Biodynamic agriculture could, if the buffalo were included therein, restore to the American soul something of its function. I intuited the great purpose Devon saw and systematically pursued.

Here is another experience: Devon brought along some of his mixture of Buffalo horn powder, and I asked him if I might have it at the end of the conference. He gave it to me right then. I put the bag in my pants pocket, but I wondered after a short time what sort of energy I was collecting there. I set the mixture near my chair, which is something I had never felt the need to do before. A few days later he asked me whether he could see the mixture again. I showed him the bag. He opened it and took a deep breath of the aroma of the mixture and said that he now felt well again. He needed the contact with his animals, which he achieved with the help of this preparation.

We also talked about our research work, exploring the super-sensory method of direct perception of the imaginative powers enclosed in our preparations. One question that arose at the time was the question about what organ is best for putting together the dandelion preparation: the mesenterium (*peritoneum*) or the large net (*omentum majus*).

We found through our investigations that the latter is the right organ, since it is directly related to the highest spirituality of the animal, very much in tune with the process by which one later produces the dandelion preparation, which is able to bring the spiritual side of the silica into contact with the earth. Devon smiled and said, "You know we can take the omentum of a male buffalo and lay it out in front of us: then we can make contact with our tribal spirit and read the past and future." Feeling the spiritual affinity between these two practically distinct activities, I said only, "that is a perfect fit." The omentum majus as door or window to the spiritual world!

Who was Devon? A modern homeless one, lacking a spiritual home, seeking a new home and finding it? I think that instead he was a messenger or mediator. He searched for and found his task in life in the double affinity with Indian and biodynamic ways. He worked with a concern for the future, aiming through a strictly individual life to serve the larger community.

But the question remains: why did he have to pass on? For myself I have found an answer: so that we may awaken and understand his purpose, individualizing it according to our own nature. Let us too connect the ancient with the future! Devon lived in a spiritual relationship with nature, culture, and especially his buffalo herd. He found that relationship in his own soul. Let us try in this way to relate our work to our soul; when I remember him, I feel his gentle smile around me, the power of his trust flowing into me.

Dear Devon, I thank you for the short but intense encounter.

Dr. Uli Johannes König *works with the Institute for Biodynamic Research in Darmstadt, Germany.*

An Unofficial Farm Report
by Anke van Leewen

DURING MY BIODYNAMIC TRAINING in Switzerland, I began working for a research project conducted by the "Section for Agriculture" at the Goetheanum in Dornach. The project was called *Biodynamic Preparations: Practices of Production and Application Worldwide*. The aim of this project is to portray the living diversity of practical knowledge around biodynamic preparations everywhere. Therefore, case studies with farmers, preparation experts or groups were being done all over the world.[†]

That is the official story about what led me to visit Devon Strong's farm as a researcher in the fall of 2015. I was there on behalf of the *Biodynamic Preparations* research project. But there is an unofficial story as well, one that started when I first heard about Devon's work. That was in May 2014, when our research group had a meeting with Uli Johannes König in Darmstadt, a well-known expert on biodynamic preparations. Uli was a fan and promoter of Devon's preparations and told us about Devon's ceremonies and communication with his buffalo, and how connected Devon was with the

---

[†] More information on this study can be found here: <www.sektion-landwirtschaft.org/Biodynamic-preparations.6565.0.html?&L=1>

spirit world. I was very excited to hear about all of that, and it was clear to me that I wanted to go and see his farm for myself.

My colleague Maja, who was also working with the research project, contacted Devon some months later, to make an appointment for our official visit, as we work in teams. But for several months it was not clear if this case study could be done at all, because Devon wasn't answering his e-mail. However, when we noticed that he had been invited to the January 2015 International Biodynamic Agriculture Conference at Dornach, Switzerland, we decided to meet him there and talk about our making a research visit to his farm. He was supposed to give a speech and lead a workshop about killing an animal at the conference. The question he was going to address was the right way to kill an animal. This question has always been important to me and goes along with my question about whether we should keep animals at all. That was one reason why I participated in his workshop, but the other one was that I wanted to get to know him.

When I first came into the room where his presentation was to take place, there were already a lot of other people there. But he noticed me coming in, and when we looked at each other, I felt something very powerful happening. I thought it was some kind of spiritual encounter, and I felt my heart was involved in some way. But I didn't think of it as being about the love between man and woman at all. I expected to see him as a teacher. Probably we were both surprised at being that struck by each other.

When we met by chance later that evening, one of the first things he said to me was: "May I give you a hug?" The next day when I encountered him, he was offering pemmican to people. Most of them refused it because of the smell. I tried a little, and I gave him a piece of alp cheese that I had with me. He gave me one of his bone feathers in exchange, the kind he was carving when his knife slipped in November 2015. We got a bit closer to each other each day of the conference, and afterwards Devon came to my house, so we could visit a farm up in the Alps together, before he had to leave.

After that we stayed in contact for the next seven months. During this time we were mainly writing e-mails, though not very regularly. Devon always said that having no expectations would be the best way to make our connection last. I also felt a strong inner link with him, and it didn't feel necessary to write very often. We both just continued to live our lives, but I also quietly prepared myself for the possibility of leaving my life here in Switzerland. Nevertheless it was not just a static condition, but at many times a difficult process. Despite his passing away in November 2015, I am still in a process with him. (Sometimes I even think that the long time we had apart prepared us, or mainly me, to have the kind of relationship we have now.)

At last the time came for my site visit to Devon's farm in the fall of 2015. The "Four Eagles Farm" comprises 220 acres of flat land and is surrounded by hills. The soil is heavy clay of volcanic origin. The climate is close to a desert climate and the rainfalls come mainly in the wintertime. Therefore the soil is completely dry in the summer and very muddy in the winter. When Devon started farming there in 2003, it was just a wide open pasture. He built a road and installed a water well himself. He told me that the first thing he did in this place was bury the horns for the horn manure preparation. In 2015 he had 17 buffalo and about thirty sheep (East Frisian milking sheep), plus three horses and some chickens. He also had a small market garden that was Demeter-certified.

I had seen pictures of the farm before, so I knew something about what to expect. But when I arrived there in the middle of September, my first impression was rather shocking. Even though the only fixed buildings at his farm were some sheds, you could lose orientation in his farmyard with its huge collection of old trailers, cars, and everything else you can imagine. There were buffalo bones, hides, and skulls everywhere too. His mobile home was also completely full, as he had been collecting anything he thought he might be able to use someday for many years. Even inside you found yourself surrounded by all kinds of animal items, a lot of skulls, snake's skin, sea otter hide, feathers, skulls, or coyote hides. He also had a primitive composting type toilet arrangement set up inside the trailer, and in many places there were open vessels full of sheep or buffalo fat. The first things he gave me to eat were some moldy bread and a piece of lamb that the coyotes had left behind. The lamb was fine, but I refused the bread. All of this showed me another side of Devon than I had seen before. He was a very complex man, who also played many important spiritual leadership roles in his community, was a biodynamic prep maker and farmer, ran his buffalo meat business from the farm and earned off-farm income from sheep shearing and horse shoeing all over the west. Among his many friends, customers, and students, his "housekeeping" was a well-known joke. He used to laughingly admit himself that he was a "hoarder."

Over the course of my visit, I found I was getting used to the way the place was. Sometimes I still wonder how I could get used to it so fast. One reason of course was that Devon did everything he could to make me feel comfortable, but it was also true that the special atmosphere of the place transcended the clutter. During our interviews for the research project, he explained that the reactions of people towards his place were a way for him to experience the effects of his preparations. A lot of people said: "What a wonderful place." Something did seem to shine out over and above the disorder of his place. Also for me it felt very peaceful, and I experienced a sense of freedom there.

For me those feelings were not only a result of his preparation work, but also related to the way Devon kept his animals. During my time in biodynamic training in Switzerland, I experienced how differently you can feel with different kinds of animals around and varied systems for keeping them. Devon didn't want his animals to feel fear. So he used no electric fences, not even when the sheep had eaten his garden twice in one year. The sheep were free to wander around the whole farm. That also means that they often had to deal with the coyotes at night themselves. His opinion was that a free and healthy animal has much more joy in life. On his radio show, he had said that more lambs may survive if you give the sheep a shelter for lambing, but the animals who escape the coyotes would be much healthier, if you don't do that. But it is also true that he usually did build a lambing shed that was often blown away during the lambing season by the powerful early winter winds in that area.

His buffalo fences also consisted only of barbed wire which was hardly a real barrier for them, as buffalo can easily jump a six foot fence or knock one a flimsy one down if they want to do it. Devon's fences were stabilized by putting willow branches in between the wires of the fencing. Since the framework of a sweat lodge is also made of willow branches, Devon had the idea that the buffalo should feel sheltered, as if they were in a big sweat lodge. During my visit they did cooperate and stayed inside.

While I was visiting the farm, one of the first things Devon and I did together was butcher a lamb. Devon's ritual for the sheep was much shorter and simpler than for the buffalo. The lamb was placed on its rump, as if for shearing. He said a prayer in Lakota and probably because of my presence, also in English. One sentence that I remember touched me deeply. He asked the individual to return to the group soul and serve the following generations. Then he placed some tobacco on the sheep's forehead. He asked the sheep to stay calm and hold its head turned away in one direction. With a fast, precise movement he cut the neck vein on the other side. Through his long experience, he knew exactly how deep he had to cut, so that he didn't hurt the trachea. The lamb was laid down on the ground and bled out within a few minutes. I asked myself if I would ever be able to do something like this. We hung up the lamb overnight on a simple tripod and cut the meat the next day.

After this experience I understood that all the bones and hides lying around still belonged to the farm, and that it would have been more odd if the whole animal was just suddenly gone. For most farmers the relationship with an animal ends when they send it to the butcher. For Devon there was no end to this relationship. He was very familiar with all the organs and animal parts, able to use nearly every bit of the animal, including bones, hides and wool. That was part of his Lakota approach to raising animals.

In biodynamic agriculture circles, people talk a lot about the worth of the ruminants, and I thought I had understood it, but I never fully grasped it myself until this experience. In the temperate climate of Germany or Switzerland (and I think it is similar in the Eastern states) which has approximately uniform moisture around the year, it is also possible to work with mulch. Organic agriculture without animals does seem to be possible there. But for the first time on Devon's farm, I could really see importance of a ruminant. In the middle of September it was still dry and hot. Sometimes there was a little rain for about half an hour, but all the moisture disappeared minutes later, and it was as dry as before. When

Devon opened the lamb up, I could see that the rumen was the single source of moist, live tissue in an almost dead external environment, where life seemed to have come to a halt. Sheltered inside of the animal, the plant lignans and cellulose get broken down in the moisture of the ruminant's four stomachs. Later in the year, when the rainfall starts coming at the end of October or beginning of November, it also starts to freeze during the nights. So the of high activity in the soil itself is very short in that climate.

Two things about Devon's way of dealing with his animals did make me wonder, because for me they didn't seem to fit with his beliefs. The first is that he wanted me to milk his sheep and create a small dairy at the farm. Devon said that he wanted the animals to have their own half-wild lives, without nursing them too much. I think that the intimate process of milking brings the animal automatically closer to the humans. In Switzerland, where I learned about farming, there are still a lot of very small herds of milking cows and some old breeds, that seem to carry a very old connection to the humans, grown through a long and hard shared life in the mountains. Most of the cows have names and in many stables you can still find a brush and a currycomb. I also think that animals that are in such a close connection to humans every day do start to get ill for that reason. They are exposed to the complicated psyche of the humans, as Devon would have put it. Often it happens, when there are conflicts in the team or changes of the staff, somehow the milking cows or other animals seem to start acting in ways that are similar to the human ones. Maybe there is some sense in it. I totally agree with Devon that breeding should happen only with healthy animals without any compromises, though when an animal is ill, I do have the urge to nurse it.

The other notable thing re animals on the farm was Devon's breeding system. He used to keep the dominant herd bull for a long time. Therefore he had to exchange the young females with other breeders to avoid incest. In his writing, he compares the family structure of the herd to a matriarchal society, because the experienced females are the leaders and make the decisions. But can there really be a matriarchal society, when all relations between sisters, mothers and daughters are broken apart in that way? I don't know how buffalo normally structure their herds, but his system seems to emphasize mainly the male part of the herd. Because of the way Devon butchered the buffalo in the midst of the whole herd, it was also necessary to have control over the herd bull and keeping the same bull for a long time was one of his ways to achieve this control.

I see a lot of similarities between Devon's way of life in general and his way of keeping animals. He relinquished all those kinds of security that we are used to having in our societies. He did hard physical work his whole life, and he was convinced that it had kept him healthy. When he had some money, he spent it, without thinking about tomorrow. I am sure that he had known fears and worries too, but in the end he followed his intuition to get what he really needed. Once he said that you need the physical, to reach the spiritual. That is, I think, is the key statement concerning his whole life and work.

Since I had been a vegetarian for several years, it sometimes seemed strange for me to be so close to someone who ate meat nearly every day and who butchered so many animals in his life. It made me question myself. Of course I felt a difference between his meat and meat I had eaten in my life. After eating his buffalo stew I felt very grounded and clear in my head. The stew was also very nutritious and satisfying, so that I didn't want chocolate or something else after dinner. It made me wonder what the worth of eating meat is? When I asked Devon, he said that he thought that it would be better for a lot of people to eat at least a small amount of meat. As was often the case, he didn't explain a lot, but kept his spiritual knowledge to himself. My own feeling now is that eating meat can give us grounding; it helps us keep our connection to the material world. Many people feel that eating meat is not "spiritual," but that feeling may also come from eating the fear and suffering of animals abused by today's industrial animal husbandry.

During my visit, it also became clear to me that Devon was preparing for a change in many ways. He had a lot of plans and visions about his farm, and he wanted it to become a diverse place run as a tribal community. Although I became part of these plans, I never got a clear feeling for the whole undertaking, despite the fact that my heart was clear. Sometimes I looked at all the mess and junk on his place and asked myself whether I could stay with him if it just went on like this. During my visit I would have really liked to start a project there, but for him, it was much more important to show me his life. Looking back now, I can see that his life and his visions were going in two different directions.

During the time I was with him, he often said that he wanted to put all parts of his life together, that everything was there but just needed to be put together. I saw the same pattern in his biodynamic work too. Until then, he had been using the preparations only in his garden and on some parts close to the farmyard itself. In 2015 he bought a field sprayer, to treat all of his land with the biodynamic preparations. The origin of Devon's initial interest in biodynamics was that he wanted to find a way to connect his organic gardening with his Lakota spirituality. He succeeded in doing that by using the buffalo parts for his preparations, but it was still a one way movement. He hadn't yet brought biodynamics back to the buffalo through applying the preparations on the whole land, which would have been probably even more powerful. Nevertheless, when I was on the farm I felt a kind of gravitational pull in the farm center itself that kept me there. But that kind of connection to his pasture land was still missing. The use of the field spray preparations would have brought everything together and created a "skin" around his whole farm.

Perhaps it is not so surprising that his favorite preparation was the nettle preparation, which is associated with the heart, but also with Mars. He used nettles for a lot of things. Just as Devon did with the conscious killing of an animal, the nettle plant can bring us to consciousness, through the sudden pain it gives us if we overlook it. A lot of people who saw the farm or pictures of it said immediately that there was too much male energy around, and that this place needed a woman. Devon always told me that it was very difficult for him to get a deer bladder for the yarrow preparation, which is associated with Venus. But in the fall of 2015 he got four of them, one from a hunter and three from roadkill animals. (I haven't heard about anybody else using roadkill deer bladders.) The yarrow preparation is also the preparation said to be able to draw things in that are needed from far away. His farm needed a lot of help from outside and a way for him to be able to focus more of his energy there.

ANKE VAN LEEWEN *was born and educated in Germany. Her major at university was biology, and in the course of her studies she became interested in biodynamic agriculture. In 2011 she moved to Switzerland for in-service training in biodynamic agriculture. Her education there included hands-on experience at different biodynamic farms and theoretical lessons about organic farming and biodynamics; she also learned many practical skills, including milking, milk processing, chain saw work, two-horse carriage driving, and herding. In 2013 she was certified as an organic farmer. In 2015 she was invited to join a research team for the Section for Agriculture at the Goetheanum, in Dornach, Switzerland.*

IV
Obituaries

Buffalo Rancher Found Dead in Knife Accident
by John Darling

DEVON STRONG, a buffalo rancher at his Four Eagles Farm in Siskiyou County and a frequent vendor at Rogue Valley Growers Market, is being remembered as a skilled teacher of animal husbandry who approached nature spiritually as part of the "sacred hoop of life." Strong, the father of two, was raised on a ranch in Nevada. He graduated from high school in Marin County and was a Community Supported Agriculture farmer at Ashland's Wellsprings before buying 80 acres near Montague, Calif., east of Yreka, to herd his bison.

When he failed to show up at scheduled meetings over the weekend before Thanksgiving, friends went to his ranch on Monday, Nov. 23, and found he had died accidentally while carving a feather from bone, the knife apparently slipping and cutting his femoral artery. He was 58.

Strong was an elder in the New Warrior organization in Ashland and served in the Boys-to-Men trainings that helped teen boys move into the responsibilities of manhood. Fellow Warrior Jack Lieshman said, "I've always felt such a deep connection to him because of his own connection to his path. He was always learning ... To me, his teaching came from who he was as a man; through his deep humility, his easy smile and warmth for all who were in his presence, his love for all life (and the deeper connections to the Mother)."

A tribute, written by Catherine Preus on *Edible Shasta-Butte* while he was alive, notes, "Long dark pony tail, muscular body, odor of sheep and the barnyard, perpetually smiling, weather-reddened round face, Devon Strong filled my ... apartment with his presence. He handed me frozen buffalo shank in two small packets neatly labeled and wrapped in white paper, then proffered a third packet. 'I brought you some London broil to try.' As he handed me the little packets in an almost reverent manner, I felt a special thankfulness for the medicine, as Devon calls it, and the sacredness of this transaction, for I knew I was participating in what Devon calls the Sacred Hoop of Life."

Strong is quoted in her article, describing the right way to kill a buffalo: "Take four days, do the ceremony this way, sing these songs, make 150 prayer ties, and use a blade. The animal will be ready for you and it will happen like this ... a bullet to the brain is the ultimate violence; it's unnecessary. When somebody shoots a gun it's a spiritual shock to the animal, it's a psychic shock to the whole region—the ripping of the air with the bullet—there's a violence with a bullet that is completely lacking when you use a blade."

Strong learned sacred Lakota ways from Wallace Black Elk, a frequent teacher and shaman in Ashland in the 1980s and 1990s, he says in the article—and, with his partner, Ami, operated the ranch on those principles. He was a Lakota singer and pipe carrier.

He is survived by a son, Zachary van Buuren of Portland. A daughter, Janneke van Buuren, died last spring in a drowning accident at Topaz Lake, California. Both graduated from Ashland High School.

JOHN DARLING *is a freelance writer in Ashland, Oregon.*

Some Thoughts on Devon Strong
by Jim Fullmer

I FIRST MET DEVON 30-odd years ago in the early days of the Oregon Biodynamic Group when we were making the preparations here on my farm in Kings Valley, Oregon.

Devon connected with Lakota spirituality and applied this to Biodynamic agriculture. A very powerful concept for Biodynamic agriculture on the North American continent. He made preparations using Bison rather than the domestic European cow. I have used his preparations on my own farm for many years and discovered that my own bovines, a motley crew of free willed Scottish Highlanders, greatly appreciated the practice. They often stood in reverence, watching as the droplets of Bison 500 plunged to the Earth's surface. My farm loved this practice too. My farm is in a coastal range drainage valley at the narrows of which a cavalry fort once stood where the tribal people of this land, the Kalapuya, were metered in and out of the valley before eventually being forced from their home here to reservations. On many levels, here and many places on this continent, damage was done.

For the North American continent the work that Devon started needs to be continued. I feel there is a deep karmic healing that somehow needs to be facilitated by such practice. When taking a bison, Devon did this by hand and as a prayer. Last year Devon's farm was Demeter certified. I had the honor of going down to visit him. It was incredible to be in the Bison herd with Devon, and also with his flock of sheep. Very calm, very connected and full of positive intention.

Ode to Devon! He did phenomenal work while walking this Earth and he's certain to continue from beyond.

JIM FULLMER *is the Co-Director of Demeter USA.*

V
Aftermath

THE STRONG BUFFALO STORY
by Craig Strong

TOUGH, UNTAMEABLE, and with energy and personalities unlike any other hooved animals, buffalo were the first grazers of the open plains. Slaughter of the great herds in the 1800s mirrors the decimation of the two-legged tribes that depended on buffalo for food, shelter, clothing, and deep relation to the land. We white people have a terrible and tragic legacy to bear.

Rather than wearing this legacy as a psychic burden, Devon Strong embraced the native Lakota traditions, learned the language and songs, and carried on his own version of deep relation to the land, including the keeping of a buffalo herd. Tribal elder Dave Chief made him an honorary son within the tribe, and he became a pipe carrier and leader of ceremony at many lodges and events.

Part of Devon's enthusiastic and rebellious abandonment of American middle-class values included organic farming, which soon led him to biodynamic farming in keeping with the teaching of Rudolf Steiner. The spiritual nature of biodynamics merged with native traditions in my brother, and he broke new ground in connecting the spirit energy of animal, soil preparations, and human relation to our mother earth.

The alternative lifestyle and creative construction at Four

Eagles Farm did not match or mix well with the cattle rancher neighbors or the state. Often the Buffalo Man found himself on the wrong side of the law when his headstrong herd were on the wrong side of their home-made fence. He always got them back (with bison lives lost on a few occasions), but fines, threat of lawsuits, and uneasy relations were ever present elements at the farm that refused to conform—to anything!

Devon spoke to his buffalo in ways beyond words, sometimes with a drum, sometimes in song and voice, and sometimes in ceremony. Kills were made on foot with a spear following four days of preparations and prayers for a shared understanding of what was to pass. Always aware that death is an inevitable part of life, the Buffalo Man made the crossing in the same way as his animals, consciousness suddenly flowing out of his body with the life blood. His only regret, most likely, being the mess left for those of us still living.

For me, a cataclysm in my heart and soul has passed through me twice, leaving me changed forever in finishing my life without my brother. Once on hearing the news, and again the following night at Four Eagles Farm, alone. Now the tears come and go at irregular intervals, appreciated for the pain, accepted for the time, and set aside to face the near impossible task of addressing the chaos left behind. No one will ever follow the path Devon took in life, and no one should try. But we must still pick up the pieces and tell the story.

The biggest and most pressing pieces were the 17 animals whose combination of strength, speed, and independence is unmatched by any other animal that humans have called "livestock." They respected the fences my brother made as much by agreement as by actual containment.

After Devon died, the buffalo knew their man was not on the farm anymore, and began "acting out" by breaking out. Our good caretaker Jeff was run ragged chasing after them and returning them to Four Eagles Farm. Fortunately, they respected Jeff and knew where home was, and so were easily pushed back into their pasture—for a while.

Meanwhile, nearby ranchers were getting irate and the risk of collision with cars was growing; the local agriculture inspector was on our case, but considerate enough of the situation not to levy hefty fines on a daily basis when the animals were on the loose. Something had to be done and soon! Calls to nearby buffalo ranching operations informed us that the easiest, and perhaps only, solution was to kill them all and harvest the meat. While we seriously considered this, especially for the older and more dangerous individuals, it was clear to all our family and those close to the ideals of Four Eagles Farm that this would be horribly dishonorable to everything my brother stood for and believed.

And so I started a fundraising site to relocate the animals; my father invested in a lot of really good feed so the buffalo might be more content to stay home; Jeff put in extra effort to find the weak spots in the fence and repair them; and the rancher neighbors met and decided to help out for everyone's benefit by trying get the buffalo into their corrals for moving. Meanwhile, halfway across the country, Lakota buffalo herdsman Ed Iron Cloud began a 1,500 mile trip out to Four Eagles Farm.

After that, the buffalo stayed in their pasture for a week while we assembled and held a very powerful private ceremony for the Buffalo Man. He was honored well in full Lakota tradition, led by tribal elder Jeff Iron Cloud.

The next day, after Ed Iron Cloud arrived following a grueling trip including breakdowns and bad weather, we reviewed ideas for moving the herd to a corral two miles away. Panels were placed and plans laid on how they would move through the corrals into the trailer. There was concern that cow calf pairs might be separated if we did not get them all in the trailer, and concerns about older animals fighting or trampling the young.

Still with many unknowns, at least we had a plan, at least there was something to try for. No one slept well that night

The buffalo knew the land well from their many breakout experiences, but what they did not expect was to be chased out of their pasture, having always been put back into the pasture by the two-leggeds. Jeff Iron Cloud's prayers for a good day were cut short by some of the buffalo getting trapped by a partially closed gate, followed by the busy whine of the rancher's quad runners trying to round up the animals as if they were cattle. I knew we were off to a wrong start. Finally the buffalo left their pasture through a different gate and a procession of buffalo and herders started up the road, with vehicles parked at all the intersections to keep them on the route to the corrals. All seemed well until the herd was near the corral, where they could not see there was an opening since there were several rows of corral fence behind the open gate. The ranchers treated them like cattle in trying to push them too quickly and the animals responded by breaking through a barbed wire fence out to rangeland beside the road. The cattle men were good on their quads however, and diligently got them back on the road, now coming from the other direction, but the same problems remained in them not seeing a way into the corral, and they panicked under pressure from the ranchers. The rest of the afternoon included several more fences torn out by the great beasts and a growing sense of failure among the people and anger among the ranchers. There was no agreement on what was to be done. By late afternoon we had given up and began a long, slow walk back to Four Eagles Farm. Everyone was cold, hungry, and discouraged by the time the herd peaceably made their way back to the land. The good elements of what appeared to be a hopeless situation were that no one was killed or hurt, the ranchers learned that these animals could not be handled as cows are, and the buffalo appeared well in spite of the fence breakouts and strange behavior of the two-leggeds around them.

There was more talk that night of trying to load the buffalo from the dilapidated round pen on the farm, of moving them to a different corral, or of butchering the animals after all. We needed Buffalo Man to show us what to do.

The next morning I drummed and spoke to the buffalo as

they munched their morning bales of sweet grass hay, with no indication of trauma from the day before. A new plan had been laid by Ed to walk the herd to a different corral. I had strong doubts, but stopped short of losing faith. The ranchers were calmer today, and everyone worked in unison, everyone had their role. About the time the herd left their pasture I realized what mine was. I walked out into the field, sacred drum in hand, and began speaking to them in a language I did not know. The rhythm changed in accordance with the herd movements, and at times it was directed at the ranchers to slow down, but mostly it was to let the buffalo know that this was right. There was no pressure, only movement. I did not know where this new corral was, and for a long time the procession was hidden from me behind a hill. I followed a path I had never been on before, and knew I could not stop drumming. A Northern Harrier hawk accompanied me across the field, and showed me the direction they were going. I knew this to be my brother, who has visited me as raptors from time to time ever since that first night at Four Eagles. When I saw the buffalo and people again on a road a half mile off I knew they could hear me, and I realized that, for the time, I had become the Buffalo Man. The beat guided them in their slow pace along the road, around a corner and into the corral. I could not see the corral, but the rhythm changed entirely when they were in, and finally I stopped and walked slowly to rejoin my kind.

The ranchers looked at me like I was crazy, but said nothing when I used the drum to stop the animals in their desperate run around the corral perimeter and get them to look at me. The corral setup was perfect for loading; when the buffalo ran through a wide chute looking for an escape they went straight into the waiting trailer, all 17 of them. Jubilation and congratulations radiated from the men, and drumming again calmed the buffalo as panels inside the trailer were closed to keep them from having too much mobility and hurting one another.

After that it was all the business of moving them. Ed had set up a checkpoint in eastern California, Diamond Mountain Ranch, which could hold the herd for legal inspections before travelling on to the Pine Ridge reservation. The weather had given us a two day break for this event, but heavy snowfall and ice would have prevented their going farther, had the legal barriers not done so already. Fortunately, the human carriers made it back to their homes.

Diamond Mountain Ranch is a production buffalo ranch, whose owners have excellent facilities for holding them. The Four Eagles herd spent the holidays there, fattening up on rich feed and rapidly eating through the financial reserves of the fundraiser for their upkeep at $10 per day per animal. The alpha bull did clear their pasture fence and made free with some of the ranch herd cows during this time, and the rancher recognized him to be good breeding stock.

After 4 years of mild (drought) winters, jet streams over the Pacific lined up a regular barrage of cold storms through December and into January of 2016. If the weather had not prevented the rest of the trip, the delayed vet check would have. Finally the inspections were done (except for the bulls,) Ed had lined up another truck-trailer rig to finish the trip, but now there was no way to pay for it. Their month-long stay consumed all of the fundraiser monies and more, the difference being made up by Strong family donations.

Shante Miranda, a young native woman who had been following the story throughout, saw this financial shortfall coming. From her own passion for the animals and her right spirit she single-handedly put together a fundraiser concert in Ashland with donated performances by Alice DiMicele, Good Shield Aguilar, and Michelle McAfee. The concert was excellent and appreciated, and just covered the $3,400 transportation costs for the herd. They made it to the nonprofit Knife Chief Buffalo Nation at the Pine Ridge Reservation on January 11, 2016. The bulls stayed at Diamond Mountain Ranch to spread Four Eagles seed in that herd, and their sale supports fencing and leases for the Buffalo Nation. Musician Good Shield continues his buffalo work in Yellowstone to maintain awareness and support the last truly wild buffalo herd.

So many of you reading this are a part of the story. I first thank you for what is nothing less than a miraculous unity in honoring Devon and what he believed in. I know he travelled far and wide in the west, shearing sheep, shoeing horses, carrying native ceremony, and helping out all kinds of people in need. I did not know what a legend he was even in life, or how many circles of people he shared his time, energy, and spirit with. Truly, he lived a biodynamic life. The energies and materials of his biodynamic preparations and buffalo horns have been donated to the southern Oregon Biodynamic Group.

Some have asked what will be the buffaloes' fate at Pine Ridge, after all this incredible effort to get them out there? I have never been to the reservation, but what Ed has told me is they will join a herd of about 80 animals on many hundreds of acres. Individuals will be taken in a way similar to what my brother did, with ceremony for the animals and for the herd, without bullets, and using all parts of the animal taken. We all die, it is the means and ways and relationships that makes the difference between an agribusiness feedlot unconscious death and being one of my brother's herd at Knife Chief Buffalo Nation.

Health and help.

CRAIG STRONG is Devon's Strong's younger brother. He is a consulting marine biologist and runs a firm, Crescent Coastal Research, which ranges up and down the west coast monitoring populations and studying the ecology of seabirds and marine habitats. The Strong boys started life in the deserts of Nevada and eastern California, where their father pursued his passion for a cowboy lifestyle. After the age of 8, Craig and Devon lived in Bolinas, where their mother worked at the Point Reyes Bird Observatory and their stepfather was a prominent seabird biologist. While the brothers took on extremely different roles in life, they crossed paths several times a year on visits or at family gatherings. They shared *an understanding of the disastrous path of corporate rule in America, but took very different courses in responding to it in their own lives. A version of this essay appeared in the journal and on the website of the Biodynamic Association.*

Afterword
by Susan C. Strong and Craig Strong, with Anke van Leewen

DEVON STRONG LIVED A PROFOUNDLY SPIRITUAL LIFE—and death. Lacking any religious upbringing, his faith grew by adopting native traditional practices which resonated with his own spiritual nature. Lakota understanding of the primal elements earth, air, fire, and water, seeing the circle of energies these elements bring to a farm, and integrating the biodynamic principles in a holistic worldview combine in the vision Devon carried.

As Devon would say, the reality of spirit is beyond words, and is "contained" by humans in ritual, songs, drumming, lodges, and personal experience. He said he could not say how spirit worked, but he knew it did. His faith was powerful in a very real and physical sense, as his buffaloes' group soul moved the animals on the farm, and as the many people in his circles were inspired by his generosity and leadership, at lodges and other spiritual gatherings.

Although the creation of this book in the present form is the result of a tragedy, we who have put it together hope that his work and thinking will be an inspiration to others. There are a number of ways that could happen. One is that those who are involved with farming or keeping animals, even outside the world of biodynamics, might adopt his approach. Anke van Leewen reminds us that Devon thought farmers could find their own rituals to prepare themselves and their animals for slaughter, even developing special labeling like that used for kosher and halal meat. The farmers in his Dornach workshop also agreed that taking the life of an animal should happen on the farm whenever it is possible. Death should take its natural place again. Unfortunately, there are a lot of obstacles in the form of laws and rules, which make a more natural way of dealing with butchering very difficult. To create the needed changes, she points out that farmers need to get together and organize.

In addition, Anke reminds us that we can discover through Devon's work the importance of the animal part for biodynamic preparations or as Uli J. König put it: "Why do we call it a dandelion preparation instead of a mesentery preparation?" She notes that a lot of preparation makers and researchers concentrate only on the plant material.

Still another way that Devon's brand of indigenous-biodynamic practice might be adopted by others would be using different kinds of native ruminants for biodynamic work. In South America for example, the llama, the hooved animal of the Andes, so essential to indigenous Andean agriculture, might serve. In other parts of the world the yak, the camel, and even reindeer offer a similar energy relationship between indigenous people and essential animals. In the course of preparing this book, we learned that the Agriculture Section at the Goetheanum in Dornach is already researching the

Afterword

variety of ways biodynamics is being practiced all over the world. In fact, as Dr. Uli Johannes König has written, Devon's work came to their attention because of their research project. There may be more of Devon's kind of work already going on in the Americas or in other parts of the world, or more that could be stimulated with the right model.

Continuing Devon's work in a more concrete way could occur if new links could be made between his buffalo at Knife Chief Buffalo Nation and biodynamic practitioners in South Dakota, in order to create preps of the general type Devon made.

Reaching out even further than the circle of current biodynamic practice and theory in the United States, Devon's work could even become of interest to a wider audience of sustainability advocates, ecologists, environmentalists, and others concerned about the future of our ecosphere on this planet. I'm thinking particularly of thinkers and activists who also urge recognition of the spiritual power of nature found in indigenous wisdom in the Americas. Authors like Thomas Moore, *The Re-Enchantment of Everyday Life*; David Abram, *Becoming Animal*; and also Michael Harner, *Cave and Cosmos*, come to mind. Many thinkers and writers who belong in this genre come at the subject from the perspective of civilization—shifting down, so to speak. Devon's unique contribution at that level is the way he derived his thinking and practice from deeply rooted, practical contact with animals and the nuts and bolts of farming, combined with the ancestral earth-based practices of indigenous American people, and the recovered wisdom of European indigenous knowledge by means of biodynamic practice.

His was a "ground up" perspective, based on the human need to eat to survive, and what that actually means in everyday life. As such, it is unique, and far more visceral than most writers in this genre. The urgent project of reintegrating our human species within the web of life from which we came and on which we depend to survive, whether we know it or not, requires us to look at the problem from every possible vantage point.

Whatever the future of this book and of Devon's practices and thinking, it has been our pleasure and privilege to help create this book. In many ways, as Anke notes, we are only beginning to understand the factors that work together in biodynamic agriculture. Devon's creative innovations show that there is much more to discover, at every level of farming and of living too.

Appendix: Previously Published Articles by Devon Strong

My Experience Raising Buffalo
by Devon Strong

There is the nature of cowboys and Indians also to consider, especially when raising bison. Our modern reality is that most western lands are private cattle ranches that used to be bison range, or are national forest land that is leased for grazing. The history is past, both good and bad, learning from it is up to us. Repeating old mistakes is a choice that needs to be moved on from, by both Indians and cowboys (and government). Seeing cows everywhere makes me think ahead to wonder if these "cowboys" can hold on to the thought that buffalo are actually easier to raise than cows *if you know that you can make a buffalo do whatever a buffalo wants to do* rather than a cowboy making a cow do what the cowboy wants.

There is a lot less management with bison, as they survive better on less, and without most of the vaccinations required in cattle. And the meat is higher in nutrition and flavor than beef. Now the hard part is that they can be herded and tagged and handled like cows, but it takes a lot more to do it cowboy way with bison than with cows, a lot more.

If you've ever seen an angry buffalo you will remember it all your days, and handling them cowboy way makes them angry. Remember, they want to do what they want to do, not what you want, so thinking about ceremony and doing it their way, I don't even have to see them to tell them what I need. To change pasture, just leave a gate open and in a day they are there. Need them back from 10 to 15-mile breakout? I do sweat lodge and ask them to return where they are safe from outside negativity. And they do, most times.

I can't explain it in modern terms, but the old native ways, which never left the reservations, like the buffalo which did, now can work like they were made to. The strangest thing I have found is that native people still do the ceremonies, but not with the connection to buffalo that they used to have. In ten years I have not discovered anyone doing ceremony for the buffalo like I use it, they use it for healing old wounds and teaching people that the ceremonies still exist. But not using them for working with the buffalo spiritually in day-to-day ways like cowboys do everyday with cows and the landscape; like Indians did when they depended on the connection for health and help. I see cowboys do some amazing work with horses and cows, and sure hear about the times it went wrong, native cowboys sure can do it that way also.

The most stories I hear with natives and buffalo are wrecks, like the ones cowboys have learned to avoid, with buffalo fences 8 to 12 feet high and corrals of guardrail and chutes like cages set in concrete, with high tensile wire and electric fences.

It seems that most folks that are raised on western ranches have a love of the land that keeps them connected with that way of life. As a youngster I looked around and as I kept finding evidence of native people on the land and I wondered how they had lived there. Asking the native people who were still there working on the ranches got me nowhere. No, they didn't know anyone doing the old ways, but they did know of a lot of the sites in the desert that had been used. Well, most anywhere there was water was a good place for people, so I found out for myself. That is the study of anthropology and archeology, which I didn't choose to follow.

Onward I went, riding and fixing fences, making hay and irrigating, chasing cows all over the country: branding, vaccinating, roping, lots of fixing fences, early mornings saddling horses, long hours and sweaty horses, dusty, rainy, windy, cold, hot. Just to be out in the sagebrush country was worth it to me, and I seemed to be able to spot an animal in distress, any animal not quite doing right, and bring them in, whether horse, cow, sheep, dog, or goat.

My aptitude for animal care led me to look at veterinary, except I didn't like dealing with sick or injured animals all the time. I liked working with healthy animals, which led to shoeing and shearing, which I still do 30 years later. Working closely with animals is what I found to be rewarding, animal science at that time was feeds and feeding on corporate level of production, feedlots and substituting toxic waste for micronutrients, and fertilizer.

Moving more into organic, I wandered into my own life of small farm management and hay. Tired of cows, I got sheep and found that multi-useful fiber and meat was doing well on small scale farms, as well as organic hay. I started market gardening, also following the native interests.

A Lakota Approach to Biodynamics

I found that the Lakota people were still doing old time ceremony. (The 1978 American Indian Freedom of Religion Act had just passed, following the Wounded Knee uprising in 1973, and native issues were getting more attention). I was invited to attend and found a spiritual match for what I was doing on the land. But still, organic gardening and native ways didn't fit well until a friend introduced me to Biodynamics, a European based spiritual gardening technique, outlined by Rudolf Steiner in his Agricultural Course in 1924.

What spiritual things did they deal with? The native ways start with the stone, water, air, and fire; and the Europeans called them gnomes, undines, fairies, and salamanders. Come to find out, all tribes have pretty much the same style of ceremonies dealing with these, just different languages; like Gnomes and Stone People, or in Lakota, Iyan Oyate.

After trying out biodynamics and finding in it the depth I was looking for, then combining it with the Lakota, I saw that the basis of biodynamics was in the preparation making. The "preps" are the basis of applying biodynamics to the farm organism. I liked the idea of dealing with the holistic outlook of interacting with the landscape on this level. It's like the feeling of riding all over the sagebrush country in all weathers that gives one a basic connection to the land, and like the actions of doing native ceremony and feeling the same connection. Inside out, it is the same. So I discovered that cowboys and Indians had some of the same reverence for the land, just one from the inside, and one from the outside.

The next big discovery was that biodynamics uses the

cow as the basic animal for making the spiritual connection through the "preps" that revitalizes the earth energies of the living organism of the farm.

As weird as that sounds, it was proven to work all over Europe and the higher levels of nutrients in plants and fruits and health in the farms was evidence that it worked. People on that continent using biodynamics had got it right and the industry of farming has taken organic standards as a basis for healthy farming.

A note on organics is that it derived from biodynamics in the 1950s when the whole farm organism of biodynamics was simplified and just a few steps were used as an easy way to make it work. The holistic process is still the standard in biodynamics but has been cut short in organic standards. The whole process of biodynamics is much more extensive. Steiner's Agricultural Course is a must read for anyone needing full explanation, and you won't get it in one read.

Now it gets interesting: I introduced bison instead of cows as the animal element in the preps. The objections were that wild animals were not to be used in the preps: the archetype of the cow being the earth grounding force with horns and hooves that provides the vessel for storing the energy. Looking at bison, however, all the points outstanding in the archetype are exemplified, especially the head and shoulders covered with hair. The wild aspect was tamed when I talked with clarity of the native ceremonies connected with the buffalo, not just historically but in my herd. The "management" of bison using traditional ceremonies is a new one for most anyone to think on, not having been done in a century. I have been looking for any tribal recognition of this aspect for years and only found it by my own practice. Relating it to other tribal bison raisers, I finally had a couple start to practice it with similar results. The national and international and regional biodynamic groups are more accepting of this than the traditional native peoples. A most unique situation, as historically native people have been overprotective of their traditions (for good reason). This is an opportunity for the traditional aspect of ceremony to be honored for its unique property of having historic connection to spiritual roots and physical realities that cannot be replicated using any modern methods.

Bison management is mostly done with heavy duty equipment and extraordinary measures to ensure compliance of these truly wondrous beasts. They instill reverence in cowboys that can do most anything with cows and it doesn't work the same with bison. They are a different animal.

Keenly aware of family roots and protective of youngsters, family units are strongest and when disturbed can break most anything in their way. Thus the saying "you can make a buffalo do what ever it wants to" is the most accurate way to deal with them. Understanding what they want is a native way of life, which is what stirred up the notion that wiping out the buffalo was the best way to control native tribes. It has worked too well, native ways have assimilated modern living and I can no longer find the people of the buffalo among native tribes. The modern buffalo ranchers are starting to realize the traditional buffalo herds in the wild had fine-tuned the family concept into tribal relationship, while the tribal people of the buffalo have had to forsake a way of life for survival in modern times. It was only in 1978 that the old ways were restored through the American Indian Freedom of Religion Act, and since then the practice has begun to "bring back the bison and sing back the swan." But from a practical viewpoint, native people have been estranged from the buffalo for more than a century.

Having the ceremonies and applying them to the buffalo in a practical sense is something I have not heard of. Through the years of working with traditional and non-traditional native people, I have seen a number of buffalo used in ceremonies where they are hauled in trailers to the site and/or picked from tribal herds and "taken" using rudimentary ceremony; a sense of loss and grief in taking these pervades the whole process.

I cannot describe the difference when I did my first buffalo ceremony. There were traditional elders and organic vegetarians involved, and when I completed the four-day event and a five-year-old buffalo bull fell to my hand thrown spear without fear or anger, it was an uplifting release of prayers and gifting of medicine for the food and wellbeing of buffalo people on both sides of the deal. The part that amazes both traditional people and myself is that I have repeated it over 60 times. Hearing that traditional people are now starting to renew these old ways even as I have done now makes it possible to breach the subject of bringing this deep sense of renewal to the land through the biodynamic use of preps

A Lakota Approach to Biodynamics

with buffalo instead of cow, to bring it to its greatest purpose. And the native peoples can use this to its best advantage and be honored in the process of restoring both bison and traditional values to agricultural use.

I realize this is a bigger step than just raising bison or even managing bison with the old ceremonies. The whole idea of using biodynamics and buffalo and native ceremonies together can be viewed with skeptical hilarity, of course, but what if it works? Those are the words I have been hearing for years about each of them singularly. "Biodynamics? Absurd! Never heard of such malarkey! Raising buffalo? Are you nuts? More trouble than it's worth! Native ceremonies? Bunch of wannabe Indians burning sage." But put them together and you have something nobody could imagine working. "No way! Can't happen!"

Oops, it has happened. It is a long road and I don't really know where it will lead, but we never do when something like this gets started. The interested parties are all involved now for a number of years and the "outside world" known as regular America is now becoming aware that things don't have to continue down the inevitable road. We can choose organic food now, alternative education, diverse culture and music. So why not try something entirely different?

But why all three? I like the fit of them together; they complement each other and raise the possibility of creating a better agricultural product with less input (biodynamics) from buffalo as medicine for the native people and the American public, as well as making the preps for biodynamic uses to re-enliven the landscape. Honoring the traditions that have connected buffalo and people for thousands of years gives Americans the sense of connection to this land we all love so much. As much or more than cowboys and Indians inhabiting the same landscape.

A version of this article first appeared in the Spring 2013 issue of STIR.

Taking Life Seriously
by Devon Strong

The reality of animals being food (medicine) for people involves killing them—by the thousands in commercial processing operations. Consciousness is a physical condition related to living, and eliminating it painlessly is the point in turning a living animal into product to be consumed.

Recognizing consciousness in animals as a continuum of group soul connections, as well as the individual who is connected to family, involves the farmer looking at why animals are on the farm. Certainly for producing income for the farm, they must reproduce and live healthy lives. In biodynamics, there is the consideration of integrating the other senses. The animals' contribution is integral to the spiritual sheath that holds life forces on the farm. The grazing animals cover the farm physically in search of food, and their harvest of plants prunes the growth and stimulates regrowth, harvesting, digesting, and returning microbes, as well as essence, to the land.

The grazing animals need a family to reproduce— bull, cows, and calves, or ram, ewes, and lambs. The conscious choices a farmer makes in selecting livestock are based on his or her lifestyle and the land's capacity to support it, while the farmer's unconscious/superconscious is also aware of the spiritual needs, leading the mind in selecting what kind of farming we do. Our animals lead a string of annual events based on the plants, controlled by weather. We contribute as little as possible, so the decision of what animals we have is crucial to the kind of life we share. The more we interfere with the natural lives of farm animals, the more we take away from their healthy instincts.

Modern farms don't have family units. Bulls are in one pasture; mother cows are separated from feedlot calves, heifers, and fat steers—all tuned to the music of selling products. These feedlots and dairies are the epitome of modern farming, but also the source of concentrating negative life forces, creating more problems than we can solve.

The farmer chooses who lives and who dies: natural selection is now his or her option. Keeping sick lambs alive, or treating injured animals, seems to be normal on most farms today. A healthy farm should not have that problem. Accidents are preventable, health is the highest achievement, and choosing to raise animals requires the knowledge and experience to take care of problems before they happen.

That is what we try to accomplish. I select mothers who raise their own young (ones that have problems or won't take care of babies have no place), as well as make minimal efforts to recover sick animals. Injuries are usually my own fault for not having repaired a dangerous situation. We keep the farm healthy by recycling, renewing the soil with biodynamic preps and compost, renewing the blood lines, harvesting the fruits of our labor, and culling the sick and old animals. As caretakers, it is our place to recognize disease and know remedies. The herd/flock also recognizes a disease in an individual, and our observation makes us aware that they are behaving differently towards an individual. Recognizing that individual and treating it is our responsibility, spending time and money to repair what is wrong is a careful decision regarding whether to save the individual or to take care of the group. In many cases, it is the group that needs care, at the cost of the individual life. If the healing is not extensive or expensive, the individual can return, but in many cases it is death that results.

There are the healthy young animals that stand tribute for our farms. To kill these animals is to create a space for the cycle to continue. It is like the stirring of the preps—inspiring and enlivening by our hands the spiritual elements of life to take responsible action in ending the individual life of an animal. My animals and I recognize the loss of the individuality and the emotional suffering at the physical level. When blood is spilled, the earth drinks it spiritually. With my thoughts (prayers), I call out to the generational lines of ancestors to witness, as the animal families recognize that this is not the end but a continuum of cycles. We recognize this transformation of matter into spirit as spirit is transformed into matter. I share this richness of life in the sacrifice of the individual to the spiritual. Without the pain and suffering of birth, we wouldn't have the same transformation at death. Minimizing that pain and suffering is not the point;

it is about celebrating the transformation with responsibility and in honor and respect of the gift of life and death. A blade has a simple purpose, as do pain and suffering, transformational tools with spiritual connotations. As farmers, we hold that transformative power to act as priests at this time—recognizing blood and guts as the reality of transformation, that moment of birth and death as a germination of spiritual action, like the spraying of preps. The butchering of an animal creates an impact of health and help from the farm to the people, which farmer/priests recognize when we take lives to offer that connection to the herd/flock.

The Buffalo ceremony is inspired by Lakota sweat lodge. After being introduced to the traditional ceremony by full-blood elder, Wallace Black Elk, in the early 1980s, I felt a direct link of spiritual power and, a dozen years later when I started raising bison, used that connection to answer a prayer/question: How do I kill a buffalo with honor and respect? I was given, by spirit, a specific ritual like it was unrolled on a scroll, or a song you repeat as needed: offer these songs; make prayer ties; four days later, with no fear and no anger, smoke the sacred pipe, and the one will be ready; and use a blade. Seventy-five buffalo later, it continues to work. I have sometimes used a gun to kill buffalo when someone brings fear or anger and I am at risk. This also happens due to scheduling, when I cannot complete it later and must do it in that timeframe.

The reason for using a blade has been made clear—when those involved are involved in the celebration of the gift of life, rather than the loss of life. When a gun is used, it relates to the loss of consciousness, a tearing of the conscious life forces, in comparison to the relatively gentle loss of consciousness when a blade is used.

Most often, I have killed my buffalo with a spear in a small corral separated from the herd. More recently, I have been doing it with a bow and arrow and with the herd surrounding the animal when it goes down. I make a heart/lung piercing once and start singing traditional honor songs until I can walk up to the animal and place the prayers on it. I use a feather fan to notify the herd that I am doing this while they stand back, with the fresh blood in their noses, and it seems to calm them. The herd is a very closely tied family, a matriarchal structure that is very evident and is reflected in old tribal cultures. Older females lead the decisions except in the breeding season, and the youngest calves are the benevolent focus of the group. Motherhood is the price of admission to the matriarchal decision-making, and the young bulls and heifers are the outer circle, closest to us two-leggeds for the making of medicine that we need. This connection is honored by the matriarchs when I do the ceremony in that, during the days before the kill, the animal I have chosen is usually separated easily and persuaded to come into the corral.

The animal in physical reality does not know what I am asking when I approach it. The action I take is like the spraying of the preps—a physical presentation of spiritual powers—and germination of this energy makes the difference in producing the gift of what the animal/group soul offers, or the sadness of what the archetype/group soul loses. That is the animal consciousness from my point of view.

We all learn, more or less, about our conscious relationship with animals through interaction with pets, and we recognize the difference in species awareness, as cats and dogs, birds and fish have such different awareness compared to our own. In biodynamic farming, the preparation sheaths hold animal energies that relate to the species as well as the organ that the sheath represents, both cosmically and spiritually. Specifically, with prep #502, stag bladder, I was much impressed with the specifics of its use. When I tried making the #500 prep with different horns from sheep and goats (male), I was dismayed that it didn't work; the material looked the same, but, when viewed with spiritual perception, it held no energy. The animal archetypes of sheep and goats are expressive, not impressive like cows—and, it turns out, like bison also, for when I made my first bison horn manure, I was excited to feel that I had an exceptional result. Realizing the different energies represented in different species is also needed in the killing of animals. Cows and bison hold a more powerful role and responsibility on earth than do the expressive wild and domestic animals. Most pets and many domestic animals are of the nervous energy and live in physical response, rather than the seasonal response of herd animals. I wonder about the wild African herds and native culture there. The Mongolian horse culture and Tibetan yaks show our human traditional development potential, as do the sacred cows in India. So we have this potential in all places of the world, to live with honor and respect with animals.

The invitation of spiritual influences does not need to be understood. Indeed, it is not understood, as it cannot be defined by rational minds. It is the heart influence, like love, that we as humans can feel that is like the spiritual energy in biodynamics.

The spiritual is invoked, the blade applied. The animal dies consciously, bleeding out, losing consciousness. The physical body/brain dissociates, and finally the heart and lungs exhaust, and it stops. The final spasms are like an attempt at physical escape, the individual's last act. Some are quiet, some kick for a couple minutes—an individual expression given through archetypal fighting or running away. Very like the first breaths and spasms of life, there is a wonderful fulfillment at the birth/death experience when honored with respect and understanding. Then we have the vessel of life, a lifeless body—the intricacies of heart/lung, muscle/bone, hair/hide and the mysteries of endocrinology and the nervous system, blood and guts.

After I have offered song and prayer ties on the horns with the eagle fan, I let the blood onto the ground (catching some for use) and get my tractor as the herd is also accepting the loss of the individual. I remove the carcass to my outside abattoir, raising the body by the hind legs, and start skinning by opening it tail to throat and across the legs. Once skinned with legs and head removed, I open the body cavity and take out the organs and guts, harvesting the prep sheaths and fat. I let the body cool for a few hours to overnight, and then cut and wrap the meat for freezing until it is sold. I only sell locally, person to person, as a commitment to keeping the circle of life connected from the animal to the consumer, as medicine for the people. I do not ship it or sell it through other handlers. This keeps the medicine intact from the animal consciousness to the human consciousness. Not many of the people who buy it realize the potential of what we do in biodynamics or in the process of making this way of life available, but we still do it because it is a way of life that provides all we need to survive as a species with the planet intact to share with future generations. I do it this way to offer health and help to the people, animals, and land.

This article originally appeared in the Spring 2015 issue of the newsletter of the of the Biodynamic Association of Northern California.

In Partnership with Animals
by Devon Strong

THE KEY IS BEING AWARE of how we act on the farm. In biodynamic agriculture we understand an animal as being a member of a group soul. Because we intervene in their family structure we take on responsibility for sick and wounded animals. The more we learn to understand the animal, acknowledge its instincts, the more are they in turn able to bestow a spiritual energy and a protective sheath around the farm. Grazing and digesting links the animal soul to the landscape. The mechanization of agriculture works against this.

Taking the life of an animal presents us with a similar challenge to that of stirring a preparation. In both cases we need to act in full consciousness, to invite the elemental world to the event in which the animal group soul and the life stream of a single animal meet. I try hard to develop a relationship with the animal group soul and I can feel how this is acknowledged by the herd. For me it is about creating a ritual that is different for each animal species. My ritual with bison was given to me during a traditional sweat lodge and is based on the ancient Lakota rites relating to the group soul. It is a four day process which begins with a prayer and a pipe of tobacco. Then come drums and cymbals to form a connection to the group soul and bring the animals into movement.

If I have to kill a sheep or a goat, I begin with an offering. The spirit is called up, the sounds draw the animal close and the animal dies consciously. The blood is allowed to flow from the bison too using a spear thrown to the animal as it raises its leg. In a reversal of the birth process, consciousness departs from the body and just as the gift of life given so is here the gift of death. The whole herd gathers and I bring the ceremony to an end. Without fear, without anger I go among the herd and tie prayer flags to the horns. Each animal, from the youngest to the oldest, then approaches and stands round the dead body.

From the 2015 Report of the Agriculture Conference at the Goetheanum, Section for Agriculture; Dornach, Switzerland.

Acknowledgements

A great many people helped us put this book together.

Mark Ross of STIR Magazine gave us permission to reprint Devon's article, "My Experience Raising Buffalo," which appeared in their Spring 2013 issue.

Suzanne Mathis McQueen steered us to Rebecca Briggs of the Biodynamic Association; Rebecca gave us permission to reprint Devon's article, "Taking Life Seriously" from their memorial blog page for him. Rebecca sent me on to Carin Fortin, of the Biodynamic Association of Northern California, which had originally published the article in their Spring 2015 newsletter. Rebecca also gave us permission to use an earlier version of Craig Strong's story entitled *The Strong Buffalo*, published in their journal. She also led us to Ambra Sedlmayr of the Agricultural Section at the Goetheanum in Dornach, Switzerland, the world center for Rudolf Steiner's work on biodynamic agriculture.

Ambra gave us permission to reprint the remarks Devon made about being "In Partnership with Animals," printed in the 2015 Report of the Agriculture Conference at the Goetheanum, Section for Agriculture, Dornach, Switzerland, (*Accompanying Animals with Dignity into the Future*), and also posted on the memorial blog page of the Biodynamic Association devoted to Devon.

Ambra gave us suggestions and contact information for both Jean-Michel Florin and Uli Johannes König, leaders of the Agricultural Section at Dornach. These gentlemen sent us warm testimonials about their experiences with Devon when he was in Dornach, Switzerland, to present at the International Biodynamic Agriculture conferences.

Renee Watkins ably translated Uli Johannes König's German account into English.

Zachary Strong van Burren contacted Catherine Preus on our behalf.

Catherine Preus, John Darling, and Jim Fulmer all graciously allowed us to reprint their articles about Devon.

Richard D. Strong, Devon's father, selected Devon's poems included here (from a large collection compiled by Hope Izabelle), and also helped edit Devon's unfinished book draft.

John Scott Legg of SteinerBooks was invaluable in helping us through the process of getting this book ready for publication.

About the Author and Editors

DEVON STRONG ran Four Eagles Farm, a member supported organic farm located outside of Yreka, CA. He was a biodynamic farmer for twenty-one years, and a bison rancher for seventeen. He was a member of the Oregon Biodynamic Group. In addition to his farm work, he traveled all over the West shoeing horses and shearing sheep. He was raised on cattle ranches in Nevada and pursued traditional Lakota studies for thirty years; he was eventually formally adopted by a Lakota family. He ran sweat lodges and served as a ceremonial leader in many different settings in his local community. Near the end of his life, his innovative merging of Lakota buffalo ceremony with biodynamic practices led to his being twice invited to lead workshops and presentations at the International Biodynamic Agriculture Conference held at the world center of Rudolf Steiner's thought, the Goetheanum, in Dornach, Switzerland.

SUSAN C. STRONG, PH.D., is a writer, editor, trainer, nonprofit strategist, and activist. After a career that included teaching at the college level at U.C. Berkeley's Rhetoric Department and also at the Communications Department of St. Mary's College, Moraga, followed by many years of working in progressive nonprofit organizations in the U.S., she founded the Metaphor Project in 1997. The mission of The Metaphor Project (www.metaphorproject.org) is to assist progressives and liberals in mainstreaming their messages by framing them as part of the ideal American story. She is the author of *Move Our Message: How to Get America's Ear*. Her articles and blogs have been published in a wide variety of venues.

DAWN VAN BUUREN is a sixth generation native Oregonian currently living in Portland, Oregon with her husband Zachary (Devon's son), their new baby, and plentiful pets. When her penchant for pedantic proofreading isn't on the forefront, Dawn often enjoys a rousing game of Scrabble, practicing permaculture in her backyard with her chickens, hand embroidery, and herbalism.